Programming with MicroPython
Embedded Programming with
Microcontrollers and Python

Nicholas H. Tollervey

Beijing · Boston · Farnham · Sebastopol · Tokyo

Programming with MicroPython

by Nicholas H. Tollervey

Printed in the United States of America.

Published by O'Reilly Media, Inc., 1005 Gravenstein Highway North, Sebastopol, CA 95472.

O'Reilly books may be purchased for educational, business, or sales promotional use. Online editions are also available for most titles (*http://oreilly.com/safari*). For more information, contact our corporate/institutional sales department: 800-998-9938 or *corporate@oreilly.com*.

Editors: Susan Conant and Jeff Bleiel
Production Editor: Colleen Cole
Copyeditor: Amanda Kersey
Proofreader: Marta Justak

Indexer: WordCo Indexing Services
Interior Designer: David Futato
Cover Designer: Karen Montgomery
Illustrator: Rebecca Demarest

October 2017:   First Edition

Revision History for the First Edition
2017-09-25:   First Release

See *http://oreilly.com/catalog/errata.csp?isbn=9781491972731* for release details.

978-1-491-97273-1

[LSI]

Table of Contents

Foreword

In late 2012, I had an idea to build a tiny, low-powered computing machine that could run scripts directly on the "bare metal" of the underlying hardware. The aim was to make it extraordinarily easy for anyone—no matter his or her background nor technical skill level—to be able to make a digital device that uses 1's and 0's to control the real world. In my particular case, I wanted an easier way to program the little robots that I was building at the time.

At first I didn't have any particular scripting language in mind for this device, but after investigating what would be a good choice for the language, and what people would like writing in and would find easy to learn, I chose Python.

I was also interested in running a Kickstarter campaign, to see if people liked the idea and to perhaps begin to build a community around the project. The first Kickstarter campaign that ran at the end of 2013 really did kick start MicroPython: it raised funds to build 3,000 pyboards upon which MicroPython would run; it brought nearly 2,000 people into the initial community; and it attracted some very talented programmers who volunteered their time and expertise to contribute to and improve Micro-Python.

My background is theoretical physics, so I approach the design and development of MicroPython from a much more academic and research-oriented point of view, compared to simply engineering a solution to a problem. I believe this has been part of the reason for the success of MicroPython. From the outset it was not obvious that MicroPython would even work, that Python could at all be shrunk down to such a small size and run on tiny microcontrollers with very little memory. I treated the creation of MicroPython like a research project and I used analysis combined with many coding tricks, ignoring a lot of software development principles, in order to just get it working.

My general philosophy with MicroPython is to remain minimal yet usable. And the name embodies this: "micro" makes reference to microcontrollers and embedded systems, but also gives the feeling of minimalism. "Python" is the language and is a very pragmatic and usable language, designed so humans can enjoy programming. On the one extreme, making everything too minimal hurts (the ultimate minimalism is nothing!), but on the other extreme, adding masses of usability features leads to over burden and excessive resource usage. These ideas about balancing minimalism with pragmatism apply not just to the design and development philosophy of Micro-Python but also to the wider computing industry and even beyond.

Since the first Kickstarter MicroPython has seen fantastic growth, and I, along with many others in the community, have put an immense amount of work into Micro-Python, on both the hardware and software fronts. Today MicroPython stays true to its original goal of making it easy for anyone to digitally control the real world, and is widening its scope and seeing use by hobbyists and professionals, in education and research, as well as commercial products. MicroPython has actually formed its own little industry (soon to be much bigger!), and it's very humbling to see one's creation grow and attract interest from so many diverse people around the world, and for so many diverse applications.

MicroPython is constantly improving, the community of users is growing every day, and the future looks very bright indeed. I hope that MicroPython continues to inspire people, continues to help people learn, and continues to be used to build and control systems, from coffee makers to space satellites. Digital computing devices need to be programmed and MicroPython provides an efficient and enjoyable way to do just that.

Nicholas has been a great proponent of MicroPython since the early days and has contributed immensely to its development, most notably from the community and education side, and was key to the development and success of the micro:bit platform. He knows MicroPython very well, particularly from the point of view of an end user, and is well qualified to teach you about it.

So dive into it, read and enjoy this fantastic initiation into the world of MicroPython, and go out and build some amazing digital devices!

— Damien P. George
Creator of MicroPython
July 2017

Preface

This is a book about MicroPython.

MicroPython is a reimplementation of Python 3 for microcontrollers and embedded systems. It's also a staggering feat of engineering to have a reimplemention of Python that works under such constrained circumstances. Why is this important? Micro-Python empowers people: it brings one of the most beautiful, easy-to-learn, and expressive programming languages to the world of embedded development.

This book is intended for anyone who wants to learn how MicroPython is used for embedded development. I expect you to already know Python,[1] but I don't expect you to be an expert. For example, much of the material found in this book works well in an educational context; so rather than being a professional programmer, you might be a teacher or autodidact.

I aim to give you the understanding, pointers, and ideas you need so you feel confident when working on your own projects.

Most importantly, I want this book to be both fun and inspiring. After reading it, I hope you have enough enthusiasm that you dive in, feet first, to making your own MicroPython-based doohickey, thingamabob, or enchanted object.

This book will provide an overview of the sorts of devices that run MicroPython, prompt you into thinking about how best to develop and execute embedded projects, examine how MicroPython uses and interacts with hardware in order to fulfil various common outcomes and behaviours, and explore idiomatic MicroPython development. It will close with suggestions for next steps.

As with any book, it contains compromises. Some may find it too technical, others not technical enough. Alternatively, some may feel I spend too much time on certain

---

[1] For example, you're comfortable with Python's syntax, understand how to define and call a function, and know how to control program flow in your code.

subjects and gloss over others. No matter the compromises I've had to make, my intention has been to write something that is easy to read and accessible while providing enough context and signposts to resources for further study. My ultimate aim is simple: to help bring more people to the remarkable technology that is MicroPython.

It's important to note that MicroPython is a relatively young project. It's an exciting time to get involved as a community evolves, tools and infrastructure are created, features are added, and bugs are fixed. Given the sharp uptake in conference talks, workshops, and community meetings, there is a lot of enthusiasm in various communities for MicroPython and its nascent potential. Python programmers are learning how to create projects with embedded devices, embedded developers are discovering how much fun it is to use Python, and educators all over the world see MicroPython as a compelling platform for teaching children how to code.

Furthermore, since MicroPython is a reimplementation of "regular" Python, it retains Python's state as a mature language with an international community of engaged and active programmers.

Conventions Used in This Book

The following typographical conventions are used in this book:

Italic
> Indicates new terms, URLs, email addresses, filenames, and file extensions.

`Constant width`
> Used for program listings, as well as within paragraphs to refer to program elements such as variable or function names, databases, data types, environment variables, statements, and keywords.

`Constant width bold`
> Shows commands or other text that should be typed literally by the user.

`Constant width italic`
> Shows text that should be replaced with user-supplied values or by values determined by context.

 This element signifies a tip or suggestion.

This element signifies a general note.

 This element indicates a warning or caution.

Using Code Examples

Supplemental material (code examples, exercises, etc.) is available for download at *https://github.com/ntoll/programming-with-micropython*.

This book is here to help you get your job done. In general, if example code is offered with this book, you may use it in your programs and documentation. You do not need to contact us for permission unless you're reproducing a significant portion of the code. For example, writing a program that uses several chunks of code from this book does not require permission. Selling or distributing a CD-ROM of examples from O'Reilly books does require permission. Answering a question by citing this book and quoting example code does not require permission. Incorporating a significant amount of example code from this book into your product's documentation does require permission.

We appreciate, but do not require, attribution. An attribution usually includes the title, author, publisher, and ISBN. For example: "*Programming with MicroPython* by Nicholas H. Tollervey (O'Reilly). Copyright 2018 Nicholas H.Tollervey, 978-1-491-97273-1".

If you feel your use of code examples falls outside fair use or the permission given above, feel free to contact us at *permissions@oreilly.com*.

O'Reilly Safari

 *Safari* (formerly Safari Books Online) is a membership-based training and reference platform for enterprise, government, educators, and individuals.

Members have access to thousands of books, training videos, Learning Paths, interactive tutorials, and curated playlists from over 250 publishers, including O'Reilly Media, Harvard Business Review, Prentice Hall Professional, Addison-Wesley Professional, Microsoft Press, Sams, Que, Peachpit Press, Adobe, Focal Press, Cisco Press,

John Wiley & Sons, Syngress, Morgan Kaufmann, IBM Redbooks, Packt, Adobe Press, FT Press, Apress, Manning, New Riders, McGraw-Hill, Jones & Bartlett, and Course Technology, among others.

For more information, please visit *http://oreilly.com/safari.*

How to Contact Us

Please address comments and questions concerning this book to the publisher:

> O'Reilly Media, Inc.
> 1005 Gravenstein Highway North
> Sebastopol, CA 95472
> 800-998-9938 (in the United States or Canada)
> 707-829-0515 (international or local)
> 707-829-0104 (fax)

We have a web page for this book, where we list errata, examples, and any additional information. You can access this page at *http://bit.ly/programming-with-micropython.*

To comment or ask technical questions about this book, send email to *bookquestions@oreilly.com.*

For more information about our books, courses, conferences, and news, see our website at *http://www.oreilly.com.*

Find us on Facebook: *http://facebook.com/oreilly*

Follow us on Twitter: *http://twitter.com/oreillymedia*

Watch us on YouTube: *http://www.youtube.com/oreillymedia*

Acknowledgments

I want to thank my reviewers who spotted many mistakes, suggested helpful ideas, and tactfully pointed out ways in which I could improve the text. It's a privilege to work with such a talented, smart, and sympathetic group of people. Thank you, Damien George, Radomir Dopieralski, Tim Golden, Kushal Das, Peter Inglesby, Michael Foord, Carlos Pereira Atencio, Roger Tollervey, Naomi Ceder, and Carol Willing.

I also want to thank Scott Shawcroft, Tony DiCola, Limor Fried, and Phillip Torrone, all of Adafruit Industries. They have demonstrated infinite patience and provided invaluable help, advice, and support for those aspects of the book relating to CircuitPython and Adafruit's line of boards capable of running Python. The open and collaborative nature of your amazing work is an inspiration to us all.

Jo Claessens, Howard Baker, and Michael Sparks also deserve thanks as the originators of the BBC micro:bit project. All of the MicroPython resources created for the micro:bit project are the work of an international community of volunteers. Special mention should be made of Dr. Mark Shannon, who has proven to be an outstanding contributor to this community (you can blame him for the speech synthesiser, among other things). Thanks are due to all of you, no matter how small or insignificant you may believe your contribution to be.

My editor at O'Reilly, Jeff Bleiel, deserves special thanks for his encouragement, advice, and suggestions. I imagine editing a book written by someone halfway around the other side of the world is akin to cat herding via weekly video calls. Jeff, you do it with great aplomb!

Without Damien George, the creator of MicroPython, none of this would be possible. In an industry where everyone has a brain the size of a planet, Damien is the Jupiter of our solar system. His extraordinary feat of miniaturization is, quite simply, an unprecedented achievement. Damien is backed by a growing online community of volunteers who support, fix, and contribute to MicroPython. Thanks to all of you who contribute and help with MicroPython. Most importantly of all, Damien is supported by Viktoriya and Nate. I want to take this opportunity (and take the liberty of speaking on behalf of the global MicroPython community) to express our collective and sincere gratitude for all that you have done and continue to do. Bonzer effort Damo, Vik, and Nate!

Last (but not least), I want to thank my wife, Mary, and our three children, Penelope, Sam, and William, for their continued love, support, and leg-pulling. Believe it or not, they are my muses, and I wouldn't get anything done without them.

## CHAPTER 1

What Is MicroPython?

MicroPython is a reimplementation of the Python programming language (*http://python.org/*) that targets microcontrollers and embedded systems.

Microcontrollers are computers shrunk onto a single, very small chip. Embedded systems are computers that function within a larger mechanical or electrical system. Embedded systems often use microcontrollers.

This book introduces, explores, and explains MicroPython through four typical yet different devices,[1] all of which have a microcontroller at their core.

Such devices are very different to other sorts of computer. Most computers contain lots of parts: memory, storage, and processing are physically separate components containing various specialist chips. They may also contain additional parts for sound, graphics, and networking capabilities. Such computers are significantly more powerful than the resource-constrained, microcontroller-based devices used in this book.

This raises two important questions:

- Why use such small, underpowered microcontroller-based devices?
- Why use Python?

Answering these questions illustrates why there is so much excitement surrounding MicroPython.

---

1 MicroPython works on many different devices. The four used in this book were chosen because they represent the diversity of choice while being exemplars of the different sorts of device that are available. If you have a different sort of embedded device running MicroPython, the general principles outlined in this book remain the same, and it should be relatively simple to adapt the code examples.

Why Micro?

Computers are amazing and seemingly magical things.

For example, it is commonplace to make video calls to the other side of the world. Yet, just 20 years ago, this was the realm of science fiction.

> *Any sufficiently advanced technology is indistinguishable from magic.*
> —Arthur C. Clarke, *Hazards of Prophecy: The Failure of Imagination*

The quote from science-fiction author Arthur C. Clarke suggests advanced technology appears as awe-inspiring magic. He forgot to mention that any sufficiently advanced technology is boring if it's everywhere. Familiarity and ubiquity banish any sense of wonder for all but the most inquisitive.

What do I mean?

I am just old enough to remember a world without computers everywhere. First encounters with computers felt like magic. For instance, when I was at university in the 1990s, I spent hours connected to multiuser, text-based virtual worlds[2] collaborating with people connected to the server from all over the globe. It felt amazing to work, in real-time, with such a diverse and far-flung group of friends. I was especially pleased to discover I could chat with my fellow users, albeit in a textual sense. Being British I always enquired about the weather and found it strangely satisfying to instantly know conditions in San Francisco, Tel Aviv, or Singapore (usually, better than the weather in London). Knowing this information and, more importantly, knowing how to get such information felt like magic.

Depending on your age, you probably experience one of two reactions to my example:

1. A tingle of nostalgia for those wonderful, exciting, and amazing early days of the internet.
2. Spare us reminiscences of the good old days; haven't you heard of social media?

Today, my own children are unsurprised when they video-call their grandparents from mobile phones. There is no longer a sense of amazement or wonder at such feats of engineering. Rather, like generations of children before them, they're more impressed, intrigued, and entertained by the magic found in stories, legends, and fairy tales.

Why?

---

2 Such text-based worlds are called MOOs (multiuser object orientation), and many are still available online. See *https://en.wikipedia.org/wiki/MOO*.

It captures their imagination, whereas the use of everyday objects is humdrum and unremarkable.

Take the world of J.K. Rowling's *Harry Potter* books—a world populated by enchanted objects, magical forces, and incantations that bestow upon the user the ability to transform the world around them. Readers are drawn into the world because they imagine, "Wouldn't it be amazing if *I* owned an invisibility cloak, flying broom, or could cast a spell that allowed me to breathe under water?" There is a sense of wonder in the magic found in Rowling's world. The same could be said of Tolkien's Middle Earth, ancient Greek legends, any number of super heroes and Force-wielding Jedi in the *Star Wars* movies. It's fun to be immersed in such magical realms—they entertain by encouraging a sense of wonder while giving imagination the freedom to wander. It's an opportunity to ask yourself, "What would *I* do if I were the magic wielding hero?"

What has this to do with microcontrollers?

Such computing devices allow programmers to ask a similar question. Imaginations are set free to roam but, just as in literature and life, there is always a tension between right and wrong, good and evil, yes or no.

How?

Everyday objects containing small, embedded microcontrollers become programmable. If they are programmable, they have agency (the state of being in action) and autonomy (the capability of choosing how to act given certain situations). Rather than being static, dead objects, they become independent and demonstrate behaviours that bring them to life. Microcontroller-based devices are small enough to be stuck, sewn, soldered, and screwed in and on to other objects, turning everyday things into programmable "enchanted" objects.[3] Most importantly, anyone with the right knowledge can "cast a spell" in code to modify the activities and behaviour of such objects.

Just as in literary worlds, embedded systems appear magical in ways that engage and inspire. Consequently, our world becomes programmable in a way that is similar to how make-believe spells control a fictional, magical world. Imagination and a sense of wonder for computing is, in a sense, restored. We regain the opportunity to ask ourselves, "What would *I* do if I could program these devices?" As Seymour Papert tells us, someone who learns to write code "both acquires a sense of mastery over a piece of the most modern and powerful technology and establishes an intense contact with some of the deepest ideas from science, from mathematics, and from the art of

---

3 We'll look at what it means for an object to be "enchanted" in greater detail later in the book, especially with reference to a framework of ideas suggested by Professor David Rose of the Massachusetts Institute of Technology (MIT).

intellectual model building". There are many examples of people inspired to build exciting hardware projects programmed with MicroPython. These encompass a diverse range of applications: robotics, sensing and reporting on the environment, satellite control, new musical instruments, art installations, counting South Atlantic krill, and quite a number of games, to name but a few.

Objects containing embedded and potentially networked microcontrollers are "enchanted" in the same way objects are in fictional, magical worlds—you simply give them instructions. Just like in magical stories, it is important to understand how such incantations in code make enchanted embedded objects work for you.

That's where Python comes into the picture.

Why Python?

Python is an easy-to-learn, widely used, and expressive programming language (see Figure 1-1). It's easy to write what you mean in Python with concise and simple code. Additionally, Python flourishes because it has a well-organised, proactive, diverse, and welcoming global community.

Figure 1-1. In 2016 Python was ranked the third most popular programming language in the world by the IEEE. Guido van Rossum (the inventor of Python) correctly points out the omission of the "Embedded" flag thanks to MicroPython.[4]

---

4 IEEE Spectrum, "The 2016 Top Programming Languages", (*http://spectrum.ieee.org/computing/software/the-2016-top-programming-languages?utm*) posted 26 Jul 2016.

There are copious educational resources available for Python developers of all levels of experience. If you are unfamiliar with Python, you'll find many free tutorials, courses, and meetups online; and, as you grow into Python, its excellent documentation and the technical support of its community will expand your horizons.

 This book will not include a Python tutorial.

There are a huge number of resources for both experienced programmers converting to Python and beginner programmers taking their first steps in code. Many of these resources are available online for free, and O'Reilly is rightly famous for the quality of its books and online video tutorials about Python. If you're a complete beginner, I highly recommend Jake VanderPlas's free *A Whirlwind Tour of Python* (*http://bit.ly/whirlwind-python*) and Allen Downey's *Think Python* (*http://bit.ly/think-python*).

A word of warning: learn Python 3 and avoid Python 2.

Python 2 is an earlier version of the language that's still widely used for legacy reasons. It won't be supported from 2020.

While Python 2 and Python 3 are very similar, and it's even possible to write Python code that works on both versions, Python 3 is under active development, contains many improvements to Python 2, and includes useful new features that'll never be available in Python 2.

MicroPython is a reimplementation of Python 3, and, for the rest of the book, it is assumed you have a basic level of understanding of Python 3, its syntax, and idioms.

MicroPython is a full reimplementation of Python 3. Apart from some differences described in the coming paragraphs, what you know about Python also applies to MicroPython. The most obvious difference between regular Python and Micro-Python is that MicroPython is designed to work under extraordinarily constrained conditions (such as just 16 kilobytes of RAM on the micro:bit).

MicroPython runs "bare-metal" directly on the hardware: there is no underlying operating system like Windows, macOS, or Linux. All the operations and services usually provided by an operating system are handled directly by MicroPython. MicroPython has complete and direct control of the hardware, so, in effect, *MicroPython is the operating system*.

Apart from a few exceptions, all of the language features of Python are also in Micro-Python. Due to the limited nature of the devices, MicroPython does not come with the full standard library by default. Rather, depending on the device, it will come with a subset of the standard library, often reimplemented in an efficient embedded-friendly manner.[5] Usually, most of the features of a module will be available, but aspects that are considered edge-cases or redundant in an embedded context will not be implemented. All versions of MicroPython come with modules for interacting with the hardware, GPIO (general-purpose input/output) pins, peripherals, and components connected via such pins.

Since MicroPython is Python 3, you get:

- Python's style of object orientation (but without metaclasses)
- Data types (like unicode strings, integers, and floating-point numbers) and data structures (like lists, sets, and dictionaries)
- The highly dynamic nature of Python objects
- Functions as first-class objects
- Exception handling (`try`, `except`, `finally`, and the standard built-in exception classes)
- Fun features like generator functions (using the `yield` keyword), generator expressions, and list comprehensions
- The new `async` and `await` keywords in the very latest versions of MicroPython
- A comprehensive number of Python's built-in functions

Because MicroPython runs on devices with different capabilities, the availability of features and modules will sometimes be different. Of course, devices with special capabilities, like WiFi, will include modules to support these features.

For example, the micro:bit comes with a cut-down version of the standard library's `random` module, but the default build for the ESP8266 board has the ultra minimalist `urandom` module[6] instead. Because of the onboard WiFi capability, the ESP8266 port comes with both the `json` and `urequests` modules, whereas the micro:bit port does not because the micro:bit has no WiFi. MicroPython on both boards comes with the universally useful `math` module.

MicroPython's comprehensive and flexible re-implementation of Python 3 raises three interesting opportunities:

1. The wider Python community have the opportunity to try embedded development using their favourite language.
2. Embedded developers who use C and other traditional embedded languages have the opportunity to take advantage of Python's ease of use and rich libraries.

---

5 The available modules, including some not in the standard library (such as `urequests`), can be found at *https://github.com/micropython/micropython-lib*.

6 Use of "u" is shorthand for the Greek letter μ (pronounced "mu"), which stands for "micro" (small) in the metric measurement system.

3. Beginner programmers have the opportunity to start with more than just "Hello, World!" They're empowered to create engaging first projects using light, sound, sensors, and networking.

Do you make Django websites, analyse data with SciPy, or present work with Jupyter notebooks? Thanks to MicroPython, your Python skills apply in the embedded world! It is very easy to buy an inexpensive microcontroller-based board, plug in peripherals such as sensors or motors, and build something fun and useful.

As a Python programmer with no embedded experience, my first steps with Micro-Python were rewarding and fun. It's remarkably satisfying to make an LED blink via MicroPython's REPL;[7] and, as this book demonstrates, it doesn't take much more effort to learn enough to build and invent interesting embedded contraptions.

If you're an experienced embedded developer, you may be asking, "What's wrong with C?" Nothing, and I'm not in the business of provoking unhelpful language wars. There are times when C is absolutely the right language and developers must use the best tool for solving the problem at hand.

However, in many situations, MicroPython outperforms C. When prototyping, MicroPython gives you many useful features that avoid the need for tedious boiler-plate code. One C programmer reported a working prototype in MicroPython in just an afternoon, when it would have previously taken a week in C. Sometimes a few simple lines of MicroPython are able to do what would require hundreds of lines of C, making such code easier to maintain.

MicroPython is often *fast enough and manages the limited resources well enough* that any speed or memory efficiencies gained by using C are dwarfed by Python's ease of use and productivity.

In any case, where C *is the right tool*, it's possible to write MicroPython modules in C and make use of them within your Python code. It's even possible to write inline assembler as MicroPython functions.

For some beginner programmers, making a computer print "Hello, World!" on a screen is a revelation. Alas, to the vast majority of people, the end result is a "so what?"

Who can blame them?

Compared to the wonders of kitten photos, video conferencing, virtual reality, and beautifully rendered, multiplayer video games set in an almost infinite galaxy, "Hello,

---

7 REPL stands for "read, evaluate, print, loop". It's what you get when you type python from the command line and see three chevrons (>>>) prompting you do some interactive Python programming.

World!" is a bit of an anticlimax, especially if you've been led to believe this is your first step to becoming some sort of Hollywood style uber-hacker.

Playing with embedded devices or making objects that feel "enchanted" helps to overcome the disillusionment usually encountered with traditional beginner programming exercises. It's more fun to program objects that emit rude sound effects on demand, animate a series of lights on a costume, send secret messages over a network, or become the heart of an autonomous robot vehicle than it is to underwhelm the world with a greeting.

Such opportunities allow Python programmers to join the embedded community, embedded developers to join the Python community, and beginner programmers to cut their teeth in the evolving embedded Python community gathering around MicroPython. Everyone is a winner.

Embedded programming is fun. Python programming is fun. But, put them together, and the fun factor is more than doubled!

Why is this important? Because community is Python's secret weapon.

Perhaps more than any other computing community, Python's has a reputation for friendliness, openness, outreach, and the giving of time to community efforts.

Python programmers (sometimes called Pythonists, Pythonistas and/or Pythonauts) are well organised, having created the Python Software Foundation (PSF) (*http:// www.python.org/psf/*) as a rallying point for the community. The PSF was created to promote, protect, and advance the Python programming language (whose intellectual property belongs to the PSF) and to support and facilitate the growth of a diverse and international community of Python programmers. This is achieved by supporting the development of the Python programming language itself, providing technical infrastructure for the wider Python community, running and supporting a large number of international Python conferences (PyCons) around the world, and the giving of grants to individuals and organisations that promote the foundation's aims.

Details about the Python language, the PSF, the grants programme, upcoming PyCons, and user groups around the world can be found at the Python language's website (*http://python.org/*). However, if you need information specific to Micro-Python, then visit the project's own website (*http://micropython.org/*).

Why does MicroPython have its own website rather than an area on the main website? It's because of the remarkable history of the project.

It's a great story and worthy of retelling here.

MicroPython Genesis

MicroPython is the creation of Damien George.

Damien is an Australian physicist who moved to the UK to work as a post-doctoral fellow at Cambridge University. His area of expertise is revealed on his home page on Cambridge University's website:

> My research interests centre around model building using extra dimensions, and the phenomenology of such models at the electroweak scale as is relevant for their testing at the Large Hadron Collider (LHC).

Damien also studied robotics as an undergraduate and was a competitor in the Robo-Cup (*http://www.robocup.org/*), an international competition for teams who build robots that play football (soccer). He designed and wrote a scripting language from scratch for such robots. It was called Hush (*http://hush.sourceforge.net/*), and the language made it easy to change the robot's artificial intelligence on the fly.

Obviously, Damien has an enquiring mind, a skill with hardware, and is extraordinarily talented. Back in 2013, he also had time on his hands.

He wondered if it would be possible to write a version of Python for microcontrollers. Given his experience with embedded systems found in robotics and an earlier project where he'd written his own embedded-C compiler from scratch, he was uniquely qualified and got to work. Some months later, he had created "for fun" a working proof of concept.

To gauge interest in and publicise his nascent project, Damien created a Kickstarter campaign (*https://www.kickstarter.com/projects/214379695/micro-python-python-for-microcontrollers*). He promised that if he managed to raise £15,000, he would finish off his implementation of Python and provide those supporters who had paid £20 or more with a suitable microcontroller-based device called the PyBoard. He did this under the auspices of "George Robotics Limited", the company he runs with his wife Viktoriya.[8]

When the campaign closed, it had raised almost £100,000.

The world wanted MicroPython!

---

8 Viktoriya's contributions to the MicroPython project have been and continue to be fundamental to its success and growth. The sacrifices, support, and efforts of non-technical partners of free software developers are often overlooked. In this case, the MicroPython Kickstarter was very much a Damien and Viktoriya team effort, with each complementing the other's skills and expertise.

Not only did Damien's campaign exceed all expectations in terms of funding, but he delivered on his promise: the PyBoard was delivered to almost 2,000 supporters in 2014.

From that initial seed, MicroPython's community has grown.

The code is released under an open source licence and hosted on GitHub (*https://github.com/micropython*), with people from all over the world contributing features, bug fixes, documentation, tools, and new ports of MicroPython to run on many different sorts of microcontroller.

But that's not the end of the story.

In 2015, the European Space Agency (ESA) (*http://www.esa.int*) funded the development of MicroPython to determine the suitability of the language for space applications. In particular, the research focused on the dynamic configuration of payloads, such as satellites, via Python. Importantly, the valuable work Damien completed to make MicroPython more robust for critical embedded systems[9] was folded back into the wider project.

It was a winning situation for everyone: MicroPython gained the potential to fly in space, and the community received improvements to the generic MicroPython implementation.

At around the same time, the British Broadcasting Corporation (BBC) (*http://bbc.co.uk/*) announced its intention to release a small computing-in-education device to be delivered to every single 11-year-old in the United Kingdom (approximately 1 million devices). The PSF became a partner because the BBC wanted Python as one of the preferred languages for the device. Another (unnamed) partner was to create the micro:bit runtime for Python, but, in April of 2015, they pulled out.

Damien came to the rescue. Fortuitously, Damien's next-door neighbour in Cambridge was the hardware engineer at ARM who designed the board for the micro:bit project. Damien's neighbour provided him with a prototype board, and he started porting MicroPython to the device.

The timing of an email from early May 2015 confirms it took Damien only a week to get MicroPython into a usable state (used with permission):

> I signed up to mbed, exported the blinky example for the mkit and got it compiling locally using a local toolchain. And then using this I managed to get MicroPython compiling and running on the mkit! There is a surprisingly large amount of room: I could enable floating point support, aribitrary precision integers, most of the Python features and a few builtin modules. The REPL works over the USB-UART with history and tab completion. It even has a working ctrl-C (meaning you can break out of an

---

9 With an emphasis on determinism of the virtual machine and memory management.

infinite loop). I implemented a basic "pyb" module with LED and Switch classes, and a delay function. So you can do something like:

```
led = pyb.LED(1)
while True:
    led.toggle()
    pyb.delay(100)
```

Thanks in part to Damien's efforts, devices capable of running MicroPython are in the hands of a million children. Most importantly, the BBC's micro:bit project (*http://microbit.org/*) has inspired others to recreate the project in their own locale [all the source code and hardware plans have been released under an open source license (*http://tech.microbit.org/*)]. For example, developers in Germany are already in the advanced stages of releasing Calliope (*http://calliope.cc/*), based upon the micro:bit and also capable of running MicroPython.

Success breeds success, and MicroPython came to the attention of the wider world, including Adafruit Industries.

Adafruit, the brainchild of Limor "ladyada" Fried, is an open source hardware company with an international reputation for creating playful devices (such as its "Feather" line of boards), supplying components (from generic electrical parts to branded lines of components such as the popular and ubiquitous NeoPixels), and creating accessible educational resources available from its website (*https://learn.adafruit.com/*) and YouTube channel (*https://www.youtube.com/adafruit*). Adafruit's MicroPython (*https://learn.adafruit.com/category/micropython*) learning resources are numerous, inspiring, and some of the best written and helpfully illustrated to be found anywhere. If you're ever stuck for inspiration, check out these resources.

One such Adafruit device is the latest version of the Circuit Playground Express (*https://www.adafruit.com/product/3333*). It runs Adafruit's customised version of MicroPython called CircuitPython and contains a remarkable number of onboard input and output peripherals such as a speaker, microphone, touch-sensitive pins, and multicolour NeoPixels, to name but a few.

Perhaps the only thing missing from this device is wireless networking for connecting to the internet. Given how the "Internet of Things" (IoT) is the buzzword *du jour*, one would be forgiven for thinking that MicroPython is missing out.[10] It is not!

The Internet of Things is a network of everyday objects. Within these everyday objects are embedded sensors, software, and connectivity to enable data exchange. It means light bulbs, toasters, refrigerators, flower pots, watches, fans, planes, trains,

---

10 Actually, the micro:bit has a beautifully simple radio that makes it perfect for IoT projects, and the Circuit Playground Express can send and receive messages via infrared light. We'll explore such capabilities further in Chapter 10.

automobiles, and even the kitchen sink could contain programmable, microcontroller-based devices. The vision of IoT is for computing devices to permeate everything around us and interconnect across the internet. Happily, there are a couple of cheap microcontrollers—the ESP8266 and ESP32—that are perfect for IoT projects. Guess what? They both run MicroPython!

This is both a grave and exciting prospect.

Grave, because of the potential for devices under the control of third parties to be monitoring and communicating about private spaces and personal aspects of our lives.

Exciting, because those of us who program can repurpose these devices or make our own embedded hardware under our own control to create interesting and useful projects.

MicroPython is at the heart of the exciting potential of networked microcontrollers because, as mentioned, it has been ported to two WiFi-capable microcontrollers, the popular ESP8266 and its replacement, the ESP32. The ESP8266 has built-in WiFi, and the ESP32 also adds additional support for Bluetooth.

Building on their success, Damien and Viktoriya ran a second Kickstarter campaign (*http://bit.ly/kickstart-micropython-on-the-ESP8266*) in 2016 with the modest aim of raising £6,000 to help support a port to the ESP8266. They were joined in this endeavour by Paul Sokolovsky, a collaborator and core MicroPython maintainer who had been making contributions to MicroPython from just after the first Kickstarter campaign. At almost £30,000 later, it was obvious Damien and Viktoriya had another success on their hands, and development got underway. Thanks to these efforts and the support of the Kickstarter backers, MicroPython runs well on the device. It is possible to write MicroPython code and build embedded projects that communicate over the internet with a board that costs less than $5. Such boards contain both the ESP8266 microcontroller and GPIO pins to which peripherals can be attached.

Development on the ESP32 port is in the advanced stages, and it is already quite usable (if you can get your hands on a board).

Dive In!

This book explores MicroPython with the aforementioned devices. They're typical of the different sorts of devices MicroPython supports, so skill and knowledge transfer to the many other supported devices should be easy.

All of them are relatively affordable, and each is different enough from the others to demonstrate that not only is MicroPython an extraordinarily flexible platform, but also that there's an interesting diversity in capability, intention, and potential use cases in the MicroPython device ecosystem. Since new ports of MicroPython and new devices are released regularly, there is a good chance the device you may be using doesn't even exist at the time of writing. Nevertheless, the principles, techniques, and knowledge found in this book are easy to adapt to new ports and devices.

The remainder of this book is organised into the following sections:

- Introductory chapters that describe the four boards: the original PyBoard, the micro:bit, Adafruit's Circuit Playground Express, and the ESP8266/ESP32 family of boards. Each chapter describes the boards, their capabilities, how to flash MicroPython onto the devices,[11] put your code on the device, connect to the Python REPL, and make an LED blink. If you can make an LED blink, all the essentials are in place.
- A chapter on thinking about embedded development: its scope, potential, opportunities, and potential pitfalls. This will primarily use David Rose's lists of attributes (mentioned in an earlier footnote) to provide a framework.
- A series of chapters relating to various aspects and capabilities of MicroPython, introduced and explained by practical examples in code that target the four devices: visual feedback, input and sensing, GPIO, networking, sound and music, and robotics. The examples are written to be extended and to act as launchpads for your own fun projects.
- A chapter on idiomatic MicroPython. Writing code in such constrained circumstances can pose unique challenges that don't usually impact regular Python development. For example, we will cover what to do when you inevitably encounter memory-related errors. We will also consider what options you have for improving the performance of MicroPython by squeezing as much performance as possible from the microcontroller.
- A conclusion containing pointers for your next steps.

This book provides you with all the knowledge you'll need to roll up your sleeves, get stuck in, and create something wonderful with MicroPython.

Most importantly, using MicroPython will be fun. Let's get started.

---

11 The term *flash* means to erase and re-write memory. When you flash MicroPython onto a device, you're loading it into the device's memory. The term *flash* originates from the practice of flashing ultraviolet light onto an EPROM memory chip in order to erase it before being reprogrammed. These days we flash via USB.

## CHAPTER 2

PyBoard

The PyBoard is the first device developed and built for MicroPython (see Figure 2-1). It can be purchased from the MicroPython website (*http://micropython.org/*). Damien and Viktoriya's company ships it to anywhere in the world.

Figure 2-1. The original PyBoard is about the size of a large postage stamp

The board connects to your PC via a micro USB cable. This connection provides two ways to interact with the device: as a USB flash drive and as a serial-based Python REPL.[1]

 Perhaps the most common problem for people new to Micro-Python is the wrong sort of micro USB cable. There are two sorts: those that provide just power (often for charging devices such mobile phones) and those that supply both power *and* data. It is this latter type of cable that you should use.

If your board powers up (you should see an LED light up), but you don't see it as a connected USB flash drive or cannot connect to the REPL, then you probably have the wrong type of lead.

This caveat applies to all devices discussed in this book.

The PyBoard contains a small filesystem as part of its flash (on-chip) memory. It also has an SD card slot with which you can add more available storage. If you access the PyBoard as flash storage, you will be able to copy files on and off the board's filesystem. If you create a Python script called main.py on the board's filesystem, MicroPython will execute this script when it starts up. As a result, scripts can run on the board without it having to be connected to a PC.

Just like normal Python, you can copy over other Python files, import them in main.py, and organize your code into appropriate modules.

Once physically connected, it is also possible to use any serial program to connect to the board to get MicroPython's REPL prompt. Type Python commands into the REPL to have them immediately evaluated and executed. This is a great way to explore the capabilities of MicroPython, experiment, and discover how things work.

Another useful side effect is that any script that may be running from the filesystem is also in scope within the REPL. Just like regular Python, by pressing CTRL-C the program will interrupt, and you will have access to all the objects used by your script. This is extraordinarily useful for debugging scripts. It's also possible to enter a special "paste" mode by pressing CTRL-E. This lets you copy and paste large chunks of text into the REPL. Press CTRL-C to cancel or CTRL-D to evaluate the pasted code and return to normal REPL usage.

---

1 If you are using Microsoft Windows, you may need to install a driver and configure things for USB serial to work. These steps are all clearly explained in the document "The care and feeding of Pythons at the Redmond Zoo" (*https://micropython.org/resources/Micro-Python-Windows-setup.pdf*).

The Hardware

The PyBoard is, by many measures, the most powerful board used in this book. It has an STM32F405RG microcontroller, 168 MHz Cortex M4 CPU with hardware floating point, 1,024 Kb flash ROM, and 192 Kb RAM.

As well as the micro USB slot and micro SD card slot, the board also has a three-axis accelerometer, real-time clock (with optional battery backup), two switches (reset and user defined), and four LEDs (red, green, yellow, and blue).

Connectivity with peripherals and other components is via the GPIO pins that run around three edges of the board.

In addition to the micro USB slot, the PyBoard can run off batteries or another power source. Connect the positive lead of the power supply to the port marked VIN, and ground it to GND. The input voltage must be between 3.6 V and 10 V.

Be careful when connecting anything to VIN, since there's no polarity protection on the PyBoard. Put simply, make sure you plug things in the right way round!

There are several peripherals avilable for the PyBoard. All can be obtained from the MicroPython site.

Later in the book, we will make use of the colour LCD display-skin that fits on top of the PyBoard. The display is a 160 x 128 pixel, 16-bit LCD screen with backlight and a resistive touch sensor covering the entire screen (see Figure 2-2).

Figure 2-2. The colour LCD display

Another peripheral used will be the audio skin. It also fits on top of the PyBoard. It has a built-in microphone and speaker, so it's possible to both record and play sounds (see Figure 2-3).

Figure 2-3. The audio skin

Other peripherals include a Bluetooth module, prototype skins (upon which it's possible to construct test circuits), and various sorts of servo motors for making things that move.

Full details of the hardware, including schematics, layout, datasheets for components on the PyBoard, and datasheets for components found on the peripherals can be found on the MicroPython website (*http://docs.micropython.org/en/latest/pyboard/ pyboard/hardware/index.html*).

Developer Setup

The PyBoard comes with MicroPython already flashed onto the device. However, this may not be the latest version available for the board, so it is always a good idea to reflash the device. It's not a hard process, and it ensures that you get access to the latest bug fixes and updates.

First, download the latest firmware from the MicroPython website (*https://micropy thon.org/download/*). Second, you need to disconnect everything from your PyBoard and connect the DFU (Device Firmware Update) pin to the 3.3 V pin. They are right next to each other, and if you have a magnifying glass, they're also labelled on the back of the board. They are highlighted in the photograph of the front of the board in Figure 2-4.

Figure 2-4. The adjacent DFU and 3.3 V pins on the front of the PyBoard

Connecting these pins is best achieved with a male-to-male jumper wire (bundles of which can be obtained from any electrical retailer).[2]

Updating the firmware on your PyBoard requires a DFU utility. There are two options: dfu-util or pydfu. Installation instructions for dfu-util for Windows, macOS, and Linux can be found on the dfu-util project's website (*http://dfu-util.sourceforge.net/*). You can download the pydfu script from MicroPython's GitHub repository (*https://github.com/micropython/micropython/blob/master/tools/pydfu.py*). The pydfu script depends upon libusb and pyusb. You also need to be an admin-level user for the DFU utilities to work properly.

Connect the PyBoard to your computer via the USB cable. For dfu-util, type the following command (the sudo part of the command is only needed for Unix-like operating systems):

```
$ sudo dfu-util --alt 0 -D firmware.dfu
```

Remember to replace the firmware.dfu file with the latest one you downloaded from the MicroPython website.

---

2 Any conductive wire that fits inside the holes will do.

If you have more than one DFU-capable device attached to your computer (such as Apple's Magic Mouse), you may get an error message. The simplest solution is to disconnect the other devices and try again.

To use pydfu to update the firmware, use the following command:

```
$ sudo python pydfu.py -u firmware.dfu
```

After the firmware has updated, disconnect the PyBoard from your computer and remove the jumper lead between the DFU and 3.3 V pins.

When the device appears as a removable USB flash drive,[3] it will, to start with, contain four files:

boot.py
Run when the device starts and sets up various configuration options.

main.py
The main script that contains your code. This is executed immediately after boot.py.

README.txt
Contains basic information about the PyBoard.

pybcdc.inf
The Windows driver to configure USB serial as described in the "Care and Feeding of Pythons at the Redmond Zoo".

Open the main.py file in your text editor. Replace the code comment you find therein with the following snippet of code:

```
import pyb
pyb.LED(4).on()
```

The pyb module contains all the functions and classes needed to work with the PyBoard's hardware. The script above simply switches on LED 4 (the blue LED).

Save and close the main.py file, then eject (or unmount) the device, just as you would do with a regular USB flash drive. Next press the RST button just below the micro USB port to reset the board. The blue LED should light up.

---

3 On some Linux distributions, you may need to mount the device yourself. Windows may try to configure the PyBoard as a serial device. Cancel this process and refer to the "Care and Feeding of Pythons at the Redmond Zoo" (*https://micropython.org/resources/Micro-Python-Windows-setup.pdf*) document to configure things manually.

Congratulations! You've just created your first MicroPython program and run it on the PyBoard!

But we're not finished yet. You need to be able to work interactively with the PyBoard via the REPL. This will allow you to type Python directly into a command prompt and have the device evaluate your code immediately.

On Windows, assuming you have set up the Windows driver for USB serial, you should use `putty.exe` (*http://bit.ly/putty-latest*) to connect to the board. Use Window's Device Manager to discover the COM port to which the PyBoard is connected. Within the putty application, click Session in the lefthand panel, click the Serial radio button, enter the COM port discovered via the Device Manager in the Serial Line box, and then click the "open" button.

If you use macOS and Linux, you have a choice of commands to connect to the REPL. I like `picocom` because it works well and is very simple to use.

> On macOS and Linux, the `picocom` command may not be installed. If this is the case, you have two options:
>
> 1. Install the command with a package manager for your OS (for example, use `brew` on macOS or `apt-get` or equivalent package manager on Linux).
> 2. Use the `screen` command instead, although this isn't as user-friendly as `picocom`. You don't need to supply the baud rate when using `screen`.
>
> To exit `screen`, type CTRL-A CTRL-\.
>
> There are actually several alternatives to `picocom`, some of which may need you to specify the baud rate of the connection. For MicroPython's REPL, this is always 115200.

On macOS, open the terminal and type:

```
$ picocom --baud=115200 /dev/tty.usbmodem*
```

On Linux, open a terminal and type:

```
$ picocom --baud=115200 /dev/ttyACM0
```

You may have to use `/dev/ttyACM1` (or a higher number) depending on what else you have connected to your machine. Correct permissions to access the ttyACM devices may be needed; for example, you may need to be in the `uucp` or `dialout` groups (or just use `sudo`).

No matter the operating system or how you connect to the REPL, you should end up seeing the three chevrons familiar to Python programmers the world over. You may need to press return, CTRL-C [interrupt], or CTRL-D [soft reboot] for the chevrons to appear if no bytes were read from the serial port during connection.

Welcome to the world of interactive programming on the REPL! It's just like the regular Python prompt; so if you already know Python, have a look around with the `dir` and `help` built-in functions.

The following session is typical:

```
>>> print("Hello, World!")
Hello, World!
>>> 1 + 1
2
>>> 7 / 5
1.4
>>> 7 // 5
1
>>> 7 % 5
2
>>> "hello".upper()
'HELLO'
>>> import pyb
>>> pyb.LED(1).on()
>>> pyb.LED(1).off()
>>> while True:
...     pyb.LED(1).toggle()
...     pyb.delay(500)
...
```

Remember to use CTRL-C to interrupt the infinite loop at the end of the example. MicroPython's REPL has built-in command history and autocomplete if you hit TAB. It's fun for exploring how the board and MicroPython work together.

In the preceding code example, the `pyb.delay` function is used to create a pause in the infinite loop, so the LED toggles every 500 milliseconds.

For historical reasons, different boards cause a pause in the execution of code in different ways.

Depending on the device you're using, Python's standard `time.sleep` function is not always available (although it is available for the PyBoard). Furthermore, while `time.sleep` uses seconds as its unit to measure duration, other equivalent functions such as `pyb.delay` on the PyBoard and `microbit.sleep` on the micro:bit use milliseconds.

Always make sure you check what unit of measurement to use!

Congratulations, you have the PyBoard set up and ready to go! Take some time to play and explore how the device and MicroPython work together. Documentation and tutorials for the PyBoard can be found at *http://docs.micropython.org*.

## CHAPTER 3

BBC micro:bit

The BBC micro:bit is a simple yet powerful computing device for beginner programmers. It is small, cheap, and easy to use. The British Broadcasting Corporation (BBC) created the device to promote digital creativity. In other words, the BBC wants to foster the skills and confidence needed for *anyone* to make cool stuff with computers!

This isn't the first time the BBC has created a computing device for beginner programmers. In the 1980s, I first learned to program on an 8-bit BBC microcomputer (see Figure 3-1). Every school in the UK was given one and, luckily for me, my father was a head-teacher (school principal).

One weekend he came home with several large boxes containing a monitor, the computer, leads, and various manuals. His intention was to learn how to use the computer in school. However, it took only half an hour before my brother and I had managed to take over and get our young hands on the device (I was eight years old).

Compared to today's computers it wasn't particularly powerful, usable, or friendly: when you turned it on, it made a "bloop-bleep" sound and displayed a blinking cursor. To my eight-year-old self, it was daring me to type something.

I believe my first ever interaction with a computer was typing HELLO, hitting ENTER, and getting the result: Mistake.

I had *absolutely no idea* why the computer thought I'd made a mistake, but I remember feeling excited to have a computer react to something I had typed. It turns out that I had made a syntax error: the computer didn't know how to make sense of my instruction. I quickly learned that computers do not speak English.

However, the BBC micro understood a simple programming language called BBC BASIC. It was a friendly sort of a language because many of the instructions to make the computer do cool stuff were English words. Furthermore, all the instructions in

BASIC were written on numbered lines, so you knew exactly where you were in the program and in what order the sequence of commands would be executed.

Here's my first BASIC program. Can you work out what it does?

```
10 CLS
20 PRINT "YOU ARE AN IDIOT"
30 GOTO 20
```

(Actually, I copied this program from an example in the manual, but I think the manual's line 20 printed something less antagonistic.)

Can you imagine how cool this felt? It was as if I knew a special sort of magic to make computers do exactly what I wanted. All I needed to do was work out the correct incantation or spell.

Figure 3-1. A BBC microcomputer from the 1980s

What do my youthful adventures with an ancient BBC microcomputer have to do with the BBC's modern micro:bit and MicroPython?

The BBC micro:bit is both a device *and an idea* that was best articulated by the leader of the team who created the original BBC micro all those years ago:

> The aim was to democratise computing. We didn't want people to be controlled by it, but to control it.
>
> —David Allen, project editor, *BBC Computer Literacy Project*

Without the profound realization that *I could make computers do interesting stuff,* I wouldn't be a programmer today.

Given this historical context and the aim of promoting digital creativity, how has the BBC designed a device to inspire, educate, and entertain today's generation of beginner programmers?

The Hardware

The device is about the size of a credit card and comes packed with exposed and labelled hardware.

The front of the device has two buttons, labelled A and B. A 5 x 5 matrix of LEDs between the buttons acts as a simple display (see Figure 3-2). The LEDs glow at nine different levels of brightness and emit red light.

Figure 3-2. The front of the BBC micro:bit

Across the bottom is an edge connector used to attach the device to other equipment. This is the *general-purpose input/output* (GPIO): it allows the device to consume input from and generate output for other components that may be attached.

The edge connector consists of several connectors called *pins*. They are so named by convention because the GPIO hardware interfaces on many other types of device are actual pins rather than an edge connector. Some of the pins are wide enough to be labelled and for an alligator clip to be attached to them. The other pins are best accessed by plugging the device into a female edge connector attached to a *breadboard* (a device onto which electronic components can be easily placed and rearranged).

The whole layout resembles a face. The buttons are eyes, the display is a nose, and the edge connector resembles a set of teeth. This anthropomorphisation of the device (making it look like a person) is intentional: it makes the device intriguing, even when it's switched off.

The back of the device reveals the various labelled components that make the device work (see Figure 3-3).

Figure 3-3. The back of the BBC micro:bit

The edge connector (labelled PINS) continues along the back. Just above it and to the left are the device's compass (an NXP/Freescale MAG3110 three-axis magnetometer sensor) and accelerometer (an NXP/Freescale MMA8652). These sensors allow the device to work out the direction the device is pointing, how it is oriented in space, and to detect gestures such as a shake or a flip.

Above these sensors is the microcontroller "brain" of the device. It is a 32-bit Nordic nRF51822 ARM Cortex-M0 running at 16 MHz, with 256 kilobytes of flash memory, 16 kilobytes of RAM, and a 2.4 GHz radio capable of Bluetooth low energy [BLE] networking. The antenna for the radio is immediately above the microprocessor (it looks like the pattern of battlements on a castle).

To the right of the antenna at the top of the device is a micro USB port (just like the one to be found on the PyBoard). Use it to connect the device into your computer via a USB data cable. To its right is another button that, when pressed, will reset and restart the device. Finally, at the top right is a socket for connecting a battery holder.

It is important to note that the Bluetooth protocol is completely ignored by Micro-Python since the software needed to make it work takes up an inordinate amount of flash memory and RAM. However, MicroPython provides a much simpler radio module, so the devices can still communicate over a very simple network.

Significantly, all the schematics for the device's hardware are available for free (*http://tech.microbit.org/*) and covered by a liberal open source license.

Why is this important?

Anyone (including you) could use the schematics to manufacture the device. If you have the skill and desire, there is nothing to stop you from updating, modifying, and enhancing the board to your own design. It's not just the hardware that is open: all the software related to the project is released under open source licenses.

Developer Setup

Just like the PyBoard, when you connect the micro:bit to your PC via USB, it shows up as a USB flash drive. It's also possible to connect to MicroPython's REPL in the same way as with the PyBoard.

Unlike the PyBoard, the usual way to get code running on the device is to create a hex file that *combines the MicroPython runtime and your code*. This hex file is flashed onto the device by copying it onto the USB flash drive. The device isn't actually a USB flash drive; it's just pretending to appear like this so the operating system on your PC allows you to copy over hex files. Once the hex file is flashed, the device restarts, and it reverts to a relatively empty USB flash drive.[1]

The micro:bit also has a very small filesystem in its flash memory (around 20k). Each time you flash the device with a new hex file, this filesystem is erased because the hex file overwrites the flash memory of the device. However, any files you copy onto the device or save from your MicroPython script will survive between reboots. Working with the filesystem requires a special tool called microfs that uses the USB serial connection to read and write to your local filesystem. We will cover how this works later in the chapter. However, if you flash the device with a hex file containing just the MicroPython runtime (rather than the usual hex file containing the runtime combined with your Python script), you could copy over a main.py file that, upon restart,

---

1 If you're using a Mac, macOS will complain that the USB flash drive wasn't properly ejected. It is quite safe to ignore this warning.

MicroPython will detect and run, just like on the PyBoard. Furthermore, if you copy other Python files onto the device, they can be imported just like in "regular" Python.

The most important point to remember is that if you re-flash the device with a new hex file, the contents of the filesystem are permanently erased.

The quickest way to start writing code for the micro:bit is via a browser-based editor.

To ensure that Python was part of the official BBC branded web offering, UK-based volunteers for the Python Software Foundation (PSF) created the browser-based editor for the project. The source code is hosted on GitHub (*https://github.com/bbcmicro bit/PythonEditor*), and you can try it out online at *http://python.microbit.org*.

To flash your code onto the micro:bit with the browser-based editor, download the hex file from your browser (there's a big "download" button that does this for you), find it on your local filesystem, and then drag it into the directory that shows up as the USB storage device when you plug in the micro:bit. Unfortunately, browsers don't let you interact directly with the device, thus causing this drag-and-drop kerfuffle.

Whilst writing the browser-based editor, the PSF tested it with lots of teachers and students in the UK. The feedback was clear: people find it convenient to use the browser-based solution, but the user experience is terrible to the extent that it negates the initial convenience.

Users asked for an editor that didn't have such problems yet was simple and convenient enough for beginner programmers. To this end, the same group of volunteer developers have written a set of Python modules and tools to make it easy to interact with the device from a laptop or regular PC. Built on top of this work is a simple code editor called Mu (Figure 3-4). This is ideal for beginner programmers.

Mu works with Windows, macOS, Linux, and on the Raspberry Pi. Installation instructions and links to download the right version of Mu for your operating system can be found on the project's website (*https://codewith.mu/*).

In Mu, to flash MicroPython code onto the device, simply press the "flash" button. Mu also connects directly to the device's REPL by autodetecting its presence. It also provides a very simple file transfer window for putting and getting files on and off the device's small filesystem.

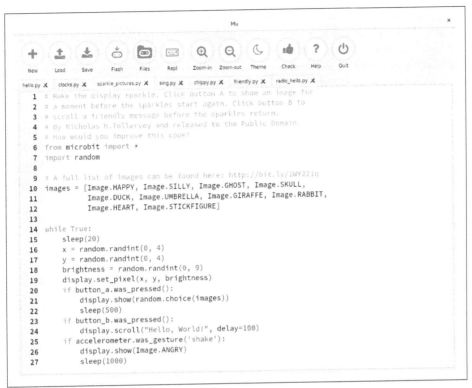

Figure 3-4. The beginner-friendly Mu editor

Mu espouses a minimalist philosophy:

Less is more.
> Mu has only the most essential features, so users are not intimidated by a baffling interface.

Keep it simple.
> It's quick and easy to learn Mu; complexity impedes a novice programmer's first steps.

Walk the path of least resistance.
> Whatever the task, there is always only one obvious way to do it with Mu.

Have fun.
> Learning should inspire fun; Mu helps learners quickly create and test working code.

If you use Mu and think, "Why doesn't Mu have feature *X*?" then you are probably advanced enough to use a *proper* programming editor, graduate away from Mu, and use the utilities created to interact directly with the micro:bit. My suggestion is, only use Mu if you need a simple and beginner-friendly programming environment. If you find you're missing features in Mu, it's easy to use the Python tools that Mu uses to interact with the device, but with your own code editor.

The two modules that Mu uses to interact with the device are called uflash (*https://uflash.readthedocs.io/en/latest/*) (for flashing the MicroPython runtime and code onto the device) and ufs (*https://microfs.readthedocs.io/en/latest/*) (for interacting with the device's filesystem). In both cases, you pronounce the "u" (μ) as "micro", as in "micro-flash". Both modules include code and command-line tools to interact with the device. Both are available as Python packages on PyPI (*https://pypi.python.org/pypi*), so if you have a relatively recent version of Python installed on your machine, use the pip command to install them in the usual way:

```
$ pip install uflash
$ pip install microfs
```

If you use the uflash command without any arguments, it will attempt to find a connected micro:bit and flash an unmodified MicroPython runtime onto the device. To flash a Python script onto the device, simply pass the .py file as the first argument to the command:

```
$ uflash my_script.py
```

Under the hood, uflash combines the default MicroPython runtime with your script so that the script runs when the micro:bit boots up. If you use the -w or --watch flag, then uflash will watch the referenced file for changes so that the device will be flashed automatically every time you change your file.

Should you have several devices plugged in, you can flash them all at once, providing you pass the paths on the filesystem where they appear as USB flash storage devices:[2]

```
$ uflash my_script.py /path/to/MICROBIT1 /path/to/MICROBIT2
```

If you're developing MicroPython and have compiled a new hex representation of the runtime, you can specify that uflash use it instead of the built-in version of Micro-Python:

```
$ uflash -r firmware.hex my_script.py
```

The -r flag could be replaced with --runtime=firmware.hex.

The ufs tool is designed to feel like FTP when interacting with the micro:bit's filesystem. For example, use the ufs ls command from your terminal to list the files on the

---

2 In this example, Unix-style paths are used. Windows paths use a \ to show the path hierarchy.

device. To delete a file on the device, you simply ufs rm *my_file*.txt. Copying a file onto the device is achieved with ufs put *path/to/my_file*.txt, whereas copying a file from the device into your current directory is ufs get *my_file*.txt.

MicroPython on the micro:bit comes with a microbit module that helps you to interact with the device directly. It also includes various other modules that make it easy to do fun things (such as make a network of devices, make sounds, and work with funky peripherals like NeoPixels). The microbit module is to the micro:bit as the pyb module is to the PyBoard.

To check everything is working, enter the following code into the browser-based editor, or (preferably) Mu, and flash the micro:bit. After a few seconds, it'll restart, and the LED in the middle of the display will blink[3] at you:

```
from microbit import display, sleep

while True:
    display.set_pixel(2, 2, 9)
    sleep(500)
    display.set_pixel(2, 2, 0)
    sleep(500)
```

If you are using Mu, it's very easy to get access to the REPL: click the REPL button! If you don't have Mu available, you can use exactly the same commands as you would with the PyBoard. For example, picocom --baud=115200 /dev/ttyACM0 should just work.

MicroPython on the micro:bit has a target audience of beginner programmers, many of whom will be teenagers. As a result, there are a number of Easter eggs in this version of MicroPython. The REPL is a good way to look for such special "features". A good place to start is to read the list of available modules found in the default text returned by the help function.

Detailed documentation for the micro:bit and a Python tutorial for beginner programmers can be found at *http://microbit-micropython.readthedocs.org/*.

---

3 Remember, the microbit.sleep function uses milliseconds to measure duration.

## CHAPTER 4

Adafruit Circuit Playground Express

Star Trek is one of my favourite science-fiction universes.

Upon reflection, a big reason for its appeal is that *Star Trek*'s fictional technology is generally a force for good. It facilitates progress (technology is used to help others), a humane and open-minded outlook (technology allows characters to live, work, and communicate with each other despite physical, physiological, and cultural differences), and fearless exploration of our universe (they fly around in spaceships!).

One of my favourite *Star Trek* technologies is the "tricorder," a device used by Mr. Spock, Bones, and others to sense the environment, make computations, and react to things with flashing lights or strange chirruping noises that obviously make perfect sense to citizens in the 23rd century.

I've often thought it'd be cool to own such a device.

With Adafruit's Circuit Playground Express, my dream has come true. Even better, it's fun to imagine Mr. Spock programming such a device in Python.

The tricorder is a classic example of an enchanted device, although not for magical reasons. The imagined technology of the 23rd century is so advanced to our primitive eyes that we react to it in the way Arthur C. Clarke suggests: it's indistinguishable from magic. The Circuit Playground Express is the antidote to such misplaced wonderment. Because it is tricorder-ish, it's packed full of sensors and modes of feedback. It lives up to its name: it's a playground for quickly learning about how embedded devices work so you too can explore strange new worlds, seek out new life and new civilizations, to boldly go where no Python programmer has gone before!

It is because of the hard work of Adafruit, who make embedded development fun and a source of wonder, that Python programmers can emulate Spock, Bones and those personnel in the red uniform that always get shot while on planet-bound expeditions.

As with the previous boards, there are two ways to interact with the device: via code stored on the onboard flash filesystem and via the REPL.

As before, when you plug the device into your computer via a micro USB cable, it'll appear as a flash storage device, and you gain access to the REPL in exactly the same way as with the PyBoard and micro:bit.

The Hardware

There are two versions of this device: an older version based upon the ATmega32u4 microprocessor that's not capable of running MicroPython, and the more powerful, newer version that uses the ATSAMD21G18 ARM Cortex M0 microprocessor. It is this latter version that we will be using in this book.

Just like a *Star Trek* tricorder, the Circuit Playground Express is packed full of input (sensing) and output (signalling) features. In Figure 4-1, if you look at the side of the board with the components on it, you'll see they're labelled and sometimes given names.

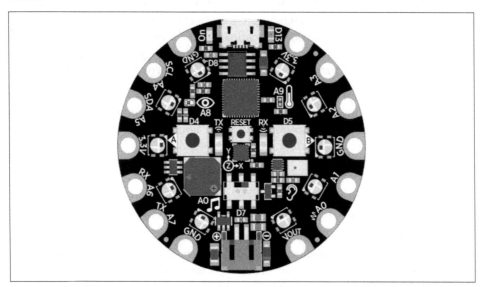

Figure 4-1. Adafruit's Circuit Playground Express is packed full of inputs and outputs.

If you hold the device with the components facing towards you and with the micro USB port at 12 o'clock, you'll notice a green LED immediately to the socket's left. Around the edges are 14 pads that make it easy to connect the device to other stuff via alligator clips. Each pad is labelled (for example, some are power-related pads labelled 3.3 V or GND). Those pads not related to power are capable of capacitive touch sensing (i.e., just like the micro:bit's pins 0, 1, and 2, they detect if they have been touched).

In the six o'clock position is the power connector into which you can provide between 3.5 V to 6.5 V DC. The board automatically regulates such voltage down to 3.3 V. The power connector works especially well with a 3 x AAA battery holder, although other options, such as lithium ion or lithium polymer batteries, could be used.

Just within the pads are 10 NeoPixel LEDs. These are extraordinarily cool since you can assign each LED an RGB colour and light it up. That's over 16 million shades of colour, so you can make blinkenlights on steroids! These NeoPixels are a great way to indicate status: red for danger, green for safety; throbbing slowly to indicate peace, flashing quickly for urgency. The NeoPixels are assigned numbers from 0–9, with position 0 to the left of the USB port and below the power LED, with the other positions counting up in an anti-clockwise direction.

In the central area of the device are three push buttons labelled A, B, and RESET. The A and B buttons are used for arbitrary on/off user interactions, whereas the reset button obviously reboots the board.

There's also a slide switch just above the battery connector (at six o'clock). It has a small nub to move to the left or right. Once in a certain position, it remains in that position (in contrast to push buttons that are only "on" when they're pressed down). Just like the buttons, the slide switch is used to indicate arbitrary on/off user input but whose state must remain persistent.

On either side of the reset button are devices for working with infrared light: a transmitter (on the left) and receiver (on the right). This allows the device to communicate with other Circuit Playground Express boards in line of sight and within range. Put simply, communication works in the same way as your TV remote control.

There are lots of environmental sensors on the board. Just above button A is an analog light sensor that can be used to measure the brightness of the ambient light. Just above button B is a *thermistor*, a sort of resistor that changes its resistance depending on the ambient temperature, thereby allowing the temperature to be measured. Between the infrared transmitter and receiver and just below the reset button is a three-axis accelerometer that allows the measurement of the gravitational force applied in X, Y and Z directions. Consequently, not only can gravity be sensed but also tilt, motion, and gestures. Below button B is a small microphone to detect audio levels by turning sound waves into electrical signals that we can, in turn, measure.

That's quite a lot of ways for the device to sense its environment!

Another component that is useful for output is just below button A. The very small black box is actually a miniature speaker. The Circuit Playground Express wouldn't be like a tricorder without the ability to make bleeps, bloops, and whistles for signalling status. It's also possible to play very simple melodies, although don't expect great audio quality, since it sounds like one of those annoying bleeping birthday cards.

Finally, in case you were wondering, the ATSAMD21G18 microcontroller is the chip just above the reset button. It runs at 48 MHz with 256 KB of onboard flash memory and 32 KB of RAM. The board also has a hefty 2 MB of additional flash memory storage.

Of all the boards covered in this book, the Circuit Playground Express has the most diverse and immediately available onboard input and output capabilities. As the name suggests, it's a great platform for playful sensing, detecting, and feedback—just like a tricorder.

Developer Setup

Adafruit uses a fork of MicroPython called CircuitPython.[1]

CircuitPython is developed as free software by Adafruit, and it welcomes contributions from the wider community. It tracks the major releases of MicroPython rather than following everything on the upstream MicroPython master branch. Work is ongoing and marked as "beta", meaning most APIs will be stable if not bug free. Adafruit makes a number of MicroPython-capable boards; and by using CircuitPython, it's able to ensure that the APIs remain consistent across its range of devices. If you learn to program the Circuit Playground Express with CircuitPython, then you'll be able to transfer your skills and knowledge to any other Adafruit board that runs CircuitPython.

The Circuit Playground Express may not come with CircuitPython flashed onto it. Also, given the ongoing development of the project, you should flash the latest version of the CircuitPython firmware onto the device in order to get the latest bug fixes and features.

It's very easy to update the Circuit Playground Express thanks to a bootloader called UF2. You simply put the device into bootloader mode then drag *.uf2* files onto the device.

At time of writing, the builds of CircuitPython for all Adafruit's boards are created as releases on GitHub (*https://github.com/adafruit/circuitpython/releases*). You'll need to download the latest UF2 version for the Circuit Playground Express.

To put the device into bootloader mode, connect it to your PC and double-tap the reset button. Once the bootloader is active, the small red LED will fade in and out, the onboard NeoPixels will turn green,[2] and the device will show up as a USB mass storage device called CPLAYBOOT.

---

1 In software, a fork is when developers take a copy of the source code from one project and start independent development that doesn't get merged back into the original project.

2 If the NeoPixels turn red, then the bootloader couldn't start. Try another USB cable.

Copy the .uf2 file onto the drive, and, once complete, restart the board. It should reappear as a USB mass storage device called CIRCUITPY.

That's it! If you encounter any problems, or you want to explore other ways to update the device, check out the comprehensive instructions on the Adafruit website (*http://bit.ly/adafruit-uf2-bootloader*).[3]

The final (optional but recommended) step is to copy the Adafruit CiruitPython drivers (*https://circuitpython.readthedocs.io/en/latest/docs/drivers.html*) bundle onto the device. It provides useful CircuitPython libraries for interacting with the hardware. The libraries are of two sorts: *foundational* (used to provide critical functionality) and *drivers* (built on top of the foundational libraries to provide access to sensors and other peripherals). The latest release of the bundle can be downloaded from the project's GitHub repository (*https://github.com/adafruit/Adafruit_CircuitPython_Bundle/releases*). Copy the contents of the downloaded ZIP file onto the device and import the modules as you would any Python library.

Just as with the PyBoard, when you plug the device into your computer, you will see it as a USB mass storage device. If you create or update the main.py file on this filesystem, the code therein will be run on startup.[4] Connect to the Python REPL on the device in exactly the same way as with the PyBoard.

The following REPL-based example makes the red LED labelled D13 toggle twice a second:[5]

```
>>> from board import D13
>>> import digitalio
>>> import time
>>> led = digitalio.DigitalInOut(D13)
>>> led.switch_to_output()
>>> while True:
...     led.value = not led.value
...     time.sleep(0.5)
...
```

As with all versions of MicroPython, use CTRL-C to interrupt the infinite loop at the end of the preceding example.

Congratulations, you have the Circuit Playground Express set up and ready to go! We'll cover how to make use of the board's many features in later chapters. The documentation for CircuitPython can be found at *https://circuitpython.readthedocs.io/*.

---

3 If you find yourself using an Adafruit device that isn't part of the Express family of boards, you'll have to use a command-line utility called bossac to update CircuitPython. This is fully explained on the referenced web page.

4 CircuitPython also allows you to use the name code.py as an alternative to main.py.

5 Remember, Python's time.sleep function uses seconds to measure duration.

If you're feeling adventurous, you could build CircuitPython from source in order to get the bleeding-edge fixes and new features.

The source code for CircuitPython is hosted on GitHub (*https://github.com/adafruit/circuitpython*) and is labelled as a fork of the main MicroPython repository. Clone the CircuitPython repository and ensure that you have the gcc-arm-none-eabi compiler installed on your machine. How you do that is down to your operating system and beyond the scope of this book, although it *should* be relatively simple if you use a package manager to install the compiler.

Assuming that you have the compiler toolchain and the source code, drop into your operating system's shell. Change the directory into the `atmel-samd` subdirectory of the repository that you just cloned. Next, use the familiar `make` command to build the firmware:

```
make BOARD=circuitplayground_express
```

The board related argument identifies the Circuit Playground Express's microprocessor as the target. If the build is a success, you'll find a `firmware.bin` file in the `build-cplay_m0_flash` subdirectory.

This file needs to be converted to the `uf2` format with a utility created by Microsoft and hosted on GitHub (*https://github.com/Microsoft/uf2/blob/master/utils/uf2conv.py*). Simply pass in the path to the `firmware.bin` file and use the `-o` flag to specify the output name. The following Linux based example is typical:

```
$ ./uf2conv.py firmware.bin -o firmware.uf2
Converting to uf2, output size: 410624,
  start address: 0x2000
Wrote 410624 bytes to firmware.uf2.
```

It is the `firmware.uf2` file that must be copied onto the device as per the earlier instructions.

## CHAPTER 5

ESP8266 / ESP32

The Internet of Things is a buzzword for embedded devices connected to the internet. Such devices are attached to all sorts of everyday objects making it easy to control or sense the object's state via the attached embedded device. It means computers are in everything around us. We are surrounded by computing devices that are uniquely identifiable and interconnected over the internet.

Such objects turn the world inside out.

In a sense, we no longer have objects that do stuff; we just have computers: computers that light things, toast bread, water plants, keep time, air-condition buildings, fly in the sky, roll along rails, and drive on roads.

Such embedded computers allow us to do three things:

Sense stuff
Collect and share data about their environment.

Control stuff
Affect changes in the environment or device.

Compute stuff
Work out what to do with data or signals such that they are autonomously useful.

As Bruce Schneier points out (*http://bit.ly/schneier-security-iot*), you can think of the sensors as the eyes and ears of the internet, the controllers as the hands and feet, and the computing power as some sort of a brain. He explains:

> *This is the classic definition of a robot. We're building a world-size robot, and we don't even realize it.*
>
> —Bruce Schneier, *Schneier on Security Blog*

At the centre of this world we find devices based upon microcontrollers like the ubiquitous ESP8266 and its replacement, the ESP32.

The ESP8266 microcontroller has WiFi and a full TCP/IP network stack built into a very cheap chip (devices can cost as little as a dollar) (see Figure 5-1). It's the creation of Shanghai-based Espressif Systems. The ESP8266's remarkable capabilities, size, and cost only came to the attention of hackers in 2014 (*https://hackaday.com/2014/08/26/ new-chip-alert-the-esp8266-wifi-module-its-5/*) and since then, much effort has been put into making MicroPython run on the device. As a result, networked Internet of Things projects are easy to build with MicroPython. The ESP8266's low cost makes the sort of networked projects that may have appeared intimidating, unfamiliar or unaffordable, a fun, cheap and accessible opportunity for more people. Python is in the Internet of Things!

Figure 5-1. A typical ESP8266-based board

Following the success of the ESP8266, Espressif Systems has released a new microcontroller with similar yet improved capabilities. The ESP32 has both integrated WiFi and dual-mode Bluetooth (i.e., it's capable of both classic and low-energy variants of Bluetooth). Furthermore, it's a dual-core microprocessor and includes cryptographic hardware acceleration for various common algorithms as well as a cryptographically secure random number generator. There is a working port of MicroPython for the ESP32, although development is ongoing. As a result the ESP8266 port of Micro-Python is currently more stable, although expect the ongoing work on the ESP32 port to mature rapidly as the board becomes more popular. The ESP8266 and ESP32 are similar enough that MicroPython makes them feel the same from a programmer's perspective. Anything you write for the ESP8266 should run with almost no changes on the ESP32.

The Hardware

Unlike the other devices mentioned in this book, ESP8266 and ESP32 are both just bare microcontroller chips rather than full, developer-friendly devices. However, you can buy different devices and development boards that contain, say, the ESP8266. This makes discussion of the hardware problematic to some extent, although there are often common features among such devices. Of course, the microcontroller's capabilities remain the same across devices.

The ESP8266[1] runs at 80 MHz and contains 64 Kb of instruction RAM and 96 Kb of data RAM. Instruction RAM is referenced during the processor's instruction fetch, whereas data RAM is used to store data. It is also capable of supporting external flash memory—although the availability of this will depend on the board.

As a WiFi-capable device, it supports the IEEE 802.11 b/g/n wireless specifications along with WEP or WPA/WPA2 authentication (as well as open networks). The microcontroller also has 16 GPIO pins for working with peripherals, although generally only 13 are available and usable.

The ESP32[2] is a far more capable device. It is a dual-core microprocessor operating at either 160 or 240 MHz, depending on configuration, and contains 520 Kb of RAM. External flash memory is also supported but will depend upon your board. The device also has 36 GPIO pins for interacting with peripherals, although generally only 34 are available and usable.

Its networking capabilities are of two types, WiFi (support for IEEE 802.11 b/g/n/e/i) and Bluetooth (version 4.2 in both classic and low-energy [BLE] modes). Security features include WFA, WPA/WPA2 and WAPI WiFi authentication, secure boot, flash encryption, and cryptographic hardware acceleration for AES, SHA-2, RSA, and ECC algorithms.

Generally, devices containing these chips will look a lot like the one in the illustration. They will probably include DIL (dual inline) headers already soldered onto the board (these look like actual pins to which it is possible to connect peripheral devices) and usually an LED or two and a couple of buttons (one of which is usually a reset button).

---

[1] See the manufacturer's data sheet (*http://bit.ly/ESP8266EX-datasheet*) for full details of the ESP8266 microcontroller.

[2] See the manufacturer's data sheet (*http://bit.ly/ESP32-datasheet*) for full details of the ESP32 microcontroller.

Two types of board are often advertised: module or development boards.

Module boards often don't have the standard connections for peripherals, so for the purposes of ease of experimentation and learning, it is probably better to purchase a development board that comes with the standard connections (such as the aforementioned DIL headers).

ESP8266 boards are remarkably cheap. They range from "white label" boards available in bulk from Chinese manufacturers for around a dollar each to branded boards (such as those produced by Adafruit) that cost around $5. Due to its relative newness and lack of availability, ESP32 boards cost around $30. However, this is likely to change both in terms of price and availability as the ESP32 becomes more popular.

Developer Setup

ESP8266

Unlike the other boards, MicroPython probably isn't already flashed onto the device, so you will have to do it yourself. It is easy to do.

You will need to download the most recent build of the firmware from MicroPython's release (*https://micropython.org/download#esp8266*) page. There are three choices: the stable build for 1024 Kb devices (the most common) and daily builds for both the 1024 Kb and 512 Kb variants of the device. You will know you have a 512 Kb variant if you get an error when flashing the device with the 1024Kb variant, "Unlikely to work as data goes beyond end of flash".

To flash the firmware, the device must be in boot-loader mode. Then use a utility to copy the firmware. Unfortunately, the precise procedure for putting a device into boot-loader mode depends upon the device, and you will need to consult the manufacturer's documentation. However, if your board has a USB connector, USB-serial converter, and the DTR and RTS pins are wired appropriately (and this is the case for most development boards such as Adafruit's HUZZAH and NodeMCU boards), then boot-loader mode will be obtained automatically if you use the `esptool.py` command.

To install the `esptool.py` utility, download the most recent release from the project's GitHub repository (*https://github.com/themadinventor/esptool/*) (making sure it is at least at version 1.2.1) or by using Python's `pip` command:

```
$ pip install esptool
```

Once installed, use the utility to erase the flash memory:

```
$ esptool.py --port PORT erase_flash
```

It is important that the `PORT` is replaced with a reference to the actual port PC to which the device is connected on your PC. For example, on Linux it will be something like `/dev/ttyUSB0`, and on Windows it will be a numbered COM port like `COM4`. This may take a few seconds, but the command will update you with its progress.

Next, flash the firmware (remembering to replace *PORT* with the actual port reference and *PATH/TO/firmware*`.bin` with the actual path to the firmware you downloaded earlier):

```
esptool.py --port PORT --baud 460800 write_flash \
    --flash_size=detect 0 PATH/TO/firmware.bin
```

If you get errors, you may need to reduce the `baud` setting from 460800 to 115200. For some boards, such as some variants of the NodeMCU board, you'll need to insert `-fm dio` after the `flash_size` argument.

If the commands run without error, then MicroPython is installed on your board. The simplest way to check is to try to connect to the REPL in the usual way (for example, `picocom --baud=115200 /dev/ttyUSB0`). Here's an example REPL session that'll blink an onboard LED on and off every half second:

```
>>> from machine import Pin
>>> import time
>>> led = Pin(2, Pin.OUT)
>>> while True:
...         led.value(not led.value())
...         time.sleep(0.5)
...
```

Unlike the other devices, the ESP8266-based device probably won't appear as a USB flash storage when you plug it into your computer, although this will depend upon your device. Nevertheless, there is a filesystem available to you, and just like all the other MicroPython boards, if you put a `main.py` file onto the filesystem, MicroPython will attempt to run this script on boot.

How do you get access to the on-device filesystem? Through the remarkable WebREPL, a means of connecting your browser to the board via your local WiFi network.

WebREPL provides two features: access to the REPL using a browser-based user interface, and the ability to upload, download, and delete files on the board's filesystem through the browser.

For this to work, the browser-based WebREPL web application will need to be open. You'll also need to set up the the board so it accepts websocket connections if you supply a password.

The WebREPL web application is conveniently hosted at MicroPython's own website (*http://micropython.org/webrepl/*), although you could run the application locally from the source code available from GitHub (*https://github.com/micropython/webrepl*).

Assuming that you point a modern browser at the web application (the MicroPython project recommends use of Firefox or Chrome), then you will see something that looks like Figure 5-2.

Figure 5-2. The WebREPL for connecting to the ESP8266.

Next, configure the password for connecting to your board. To do this, plug the device into your computer, and connect to MicroPython's REPL in the usual manner. Once you see the prompt, import the `webrepl_setup` module and follow the instructions. They'll look something like this:

```
>>> import webrepl_setup
WebREPL daemon auto-start status: disabled

Would you like to (E)nable or (D)isable it running on boot?
(Empty line to quit)
> E
To enable WebREPL, you must set password for it
New password: password
```

```
Confirm password: password
Changes will be activated after reboot
Would you like to reboot now? (y/n)
```

When this is done for the first time, you will be asked if you want to reboot the device so your changes can take effect. Upon restart and if already connected to the REPL, you will notice a message that says something like, WebREPL daemon started on ws://192.168.4.1:8266.

Return to your computer and scan for WiFi networks. You will find a new access point called something like MicroPython-020436. The default password for this network is micropythoN (note the capital "N").

Connect to this network.

With the WebREPL having already been loaded before you switched networks, copy the WebREPL daemon's URL (the one starting with *ws://*) into the box at the top left of the WebREPL application.

Immediately to the right of where you pasted the URL is a "connect" button. Click it, and a password prompt will appear in the main area of the WebREPL. Use the password you entered as part of the webrepl_setup procedure, and you should be greeted by the familiar chevrons of the Python REPL.

At this point, you will also be able to use the web forms on the righthand side of the browser application to upload, delete, and get the content of files found on the device's filesystem.

As will be seen later, it is also possible to configure the board to connect to another wireless access point. The board remembers this information and will attempt to reconnect to the last access point it was previously connected to. Once connected, it will be possible to connect to the WebREPL via the LAN rather than directly to the device acting as its own access point.

Detailed documentation for the ESP8266 port of MicroPython can be found at *https://docs.micropython.org/en/latest/esp8266/*.

ESP32

Because of the relatively new nature of the ESP32-based boards and immature (although working) MicroPython implementation for them, the flashing of firmware is rather more involved. The end result is (currently) stable but undergoing revisions. It is for this reason that I will *not* use ESP32-based boards in the rest of the book. However, the ESP32's API is designed to work just like the ESP8266's, so the examples in this book should work on the ESP32 without any modification.

Nevertheless, for those of you who feel brave and who want to live at the bleeding edge, I've included the following pointers. I deliberately use the word "pointers" since the current relatively complicated series of steps to get MicroPython running on a board will certainly become much simpler in a very short period of time.

You have been warned!

Those of you who have not skipped this section, well done; you are going to have some fun compiling the firmware yourself. To do this, clone the code from the ESP32 port of MicroPython (*https://github.com/micropython/micropython-esp32*). The most up-to-date compilation instructions are contained within the README.md file in the repository's esp32 subdirectory.

To build the firmware, you need to install the cross-compiler that targets the type of CPU found on the ESP32 (Xtensa) and download the Espressif IDF (IoT Development Framework) that MicroPython uses to work with the board. The instructions for how to do this are found in the esp-idf repository on Github (*https://github.com/espressif/esp-idf*). You only need to complete the first two steps in Espressif's instructions to meet MicroPython's requirememts.

Because Espressif's own code base is also undergoing changes, it is important you have a version that works with MicroPython. If you look in the Makefile in Micro-Python's esp32 subdirectory, you will find a line that specifies the ESPIDF_SUPHASH. Copy the referenced hash and reset the Espressif IDF repository to the required version with the following command:

```
$ git checkout <Current value of ESPIDF_SUPHASH>
```

If this step is forgotten and a version of the Espressif IDF that is not compatible with MicroPython is used, you'll get a warning when you try to build the code.

The final step for meeting requirements is setting the ESPIDF environment variable to point to the root of the Espressif IDF repository. The MicroPython documentation recommends creating a new file in the esp32 directory called makefile (or GNUmakefile if your filesystem is case insensitive) containing the following:

```
ESPIDF = path-to-espressif-idf-repository
PORT = /dev/ttyUSB0
FLASH_MODE = qio
FLASH_SIZE = 4MB
#CROSS_COMPILE = xtensa-esp32-elf-

include Makefile
```

Replace path-to-espressif-idf-repository with the correct path to the repository. Use $(HOME) rather than a tilde if you need to reference your home directory. Make sure the PORT value points to the port connecting your computer to the ESP32 device and be aware that sometimes the setting for FLASH_MODE needs to be dio rather than qio. If you didn't add the Xtensa cross-compiler to your path, use the CROSS_COMPILE setting to set its location.

With the requirements met, there are two further steps to build MicroPython. The first is to pre-compile some of the built-in scripts to bytecode with the following command executed in the root of the MicroPython repository:

```
$ make -C mpy-cross
```

The second step requires that you navigate into the esp32 directory and run make:

```
$ cd esp32
$ make
```

At last, you will find a firmware image in the build directory. Make sure your ESP32 is in bootloader mode (check with your manufacturer's documentation to see how this is done if it's not done automatically) and ensure that the port and flash settings are correct in the makefile just described.

Start by erasing the flash completely:

```
$ make erase
```

Then flash MicroPython with the following:

```
$ make deploy
```

Once finished, connect to MicroPython's REPL in exactly the same way as the other devices. But please remember, this is a relatively new microcontroller and brand-new port of MicroPython, so this is a work in progress, and not everything will be available.

This completes a quick tour of the devices that will be used in this book.

Before diving into code, I want to use the next chapter to pause and reflect upon how one might think about, explore and solve problems with embedded devices that run MicroPython.

## CHAPTER 6

Thinking Embedded

This chapter helps you think about generating embedded solutions to real-world problems.

Why is this important?

While microcontroller-based devices have been around for a long time, it is only recently that they have received broad attention due to the hype surrounding the Internet of Things (IoT). Many people have become interested in the opportunities available in this field, especially given some of the bombastic language in the media. Apparently, we are "standing at the precipice of the next transformative development, a world in which innovation becomes more human", and where "technology will be embedded in hundreds of everyday objects we already use—our cars, wallets, watches, umbrellas, even our trash cans."[1]

The antidote to such tiresome hype is reflective, grounded, and critical thinking. Such an outlook is essential for identifying and evaluating opportunities to solve important problems with genuinely useful embedded solutions.

In *The Hitchhiker's Guide to the Galaxy* series of books, Douglas Adams beautifully lampoons what happens in the absence of such thinking. He describes the Sirius Cybernetics Corporation (who manufacture all manner of everyday objects containing "advanced technology") in the following way:

---

[1] Taken from the publisher's back-cover blurb for David Rose's book, *Enchanted Objects*. As we shall see, the book is excellent and much better than the blurb suggests. Never judge a book by its cover!

*It is very easy to be blinded to the essential uselessness of [their products] by the sense of ach-
ievement you get from getting them to work at all. In other words—and this is the rock solid
principle on which the whole of the Corporation's Galaxy-wide success is founded—their fundamental design flaws are completely hidden by their superficial design flaws.*

—Douglas Adams, *So Long, and Thanks for All the Fish*

Such misapplication of inept technology is a recurring theme for Adams. For example, doors with built-in personalities not only open and close, but thank people for
using them and "sigh with the satisfaction of a job well done". They are universally
loathed by the other characters in the book.

The absence of critical thinking and a blind enthusiasm for technical solutions (for
the sake of technical solutions) is evidence of a complete lack of sympathy for the
needs of others. Worse still, it pollutes our world with annoying, incomprehensible
and often useless gizmos.

I believe a far better approach to project development is to put yourself in another's
shoes or, at the very least, listen to people and help them to adapt technology to their
own needs. After all, are you sure you know exactly what others want or need? Only
by listening, reflecting, evaluating, and experimenting is such valuable knowledge
revealed. Only then will you know what important problems to solve.

What does this entail?

Celebrate difference and diversity because it exercises our capacity for empathy and
understanding of others' lives, problems, and interests. Admit, evaluate, and learn
from mistakes, for how else are you going to adapt to change or the revelations of
new information? Keep an open mind; after all, who wants to be a prejudiced bigot
who can't see a solution because of misinformed blind spots? Look outwards, because
progress is only made through fearless exploration.

How does this relate to MicroPython?

MicroPython makes it easy to iterate, adapt, and change things, since Python pro-
motes dynamic, simple, and clear code. It gives you the flexibility to experiment, evaluate and enhance your embedded project faster. Furthermore, the wider Python
community is a diverse bunch who value the aforementioned attributes so necessary
to avoid becoming like the designers at Sirius Cybernetics.

If your project is bespoke to your own very specific need, these attributes still apply.
While you may know what you want, expanding your horizons to explore how others
have solved similar problems is both a useful and fun activity.

A large part of MicroPython's rise and popularity is because it is an amazing feat of
embedded engineering, and Python, as a language, has great features. But Micro-
Python's rise and popularity is also because its technical attributes promote an agile
and user-focused way of embedded development.

As discussed in the introduction, MicroPython also empowers three groups of people with new opportunities:

1. Python programmers transfer their experience, skills, and expertise to embedded devices, giving them a way to take part in the microcontroller/IoT movement,
2. Embedded developers enjoy the benefits of MicroPython as a platform for easy, simple, and rapid development,
3. Beginner programmers get an easy-to-use and compelling platform for learning about programming.

Such opportunities mean there are large numbers of people who are either unfamiliar with embedded devices, unfamiliar with Python, or unfamiliar with both. In all cases, many are left wondering what to build with MicroPython and the devices it supports.

For example, if you're a Python developer familiar with Django and web development, you already have the experience and technical skills needed to imagine and implement new web-based projects and opportunities. Similarly, in order to use MicroPython, you will need to develop an appreciation of the value and application of embedded devices. Eventually, you will intuitively see opportunities for solving problems by enhancing physical objects with programmable embedded devices, in the same way you perhaps already do in the context of the web.

Alternatively, if you are an embedded developer who already understands the capabilities of microcontroller-based devices, you will still need to learn how MicroPython quickly facilitates a working solution. You may be surprised by how expressive Python is compared to other programming languages commonly used for embedded development. You will also encounter the capacity of the existing global Python community to generate useful free software solutions and the necessary momentum to support them.

As a beginner programmer you will discover that Python is easy to learn; that the community is very supportive of new programmers; and that microcontroller-based devices give ample opportunities for fun, educational projects.

The fundamental questions to ask are:

- What problems can I solve?
- How can I solve them with MicroPython?
- Are embedded devices the best solution for such problems?

This chapter helps you answer the first question. Given the technical capabilities at your disposal, it will help you imagine valuable solutions, think about problems relating to how users react to and interact with embedded devices, and understand how they may use such devices to improve their lives.

The remaining chapters of the book will deal with the practicalities of the second question.

Only time will reveal the answer to the third question. Crucially, this can only happen if we exhibit reflective, grounded and critical thinking and try, experiment and learn from our mistakes in order to adjust and improve our embedded projects. A process for which MicroPython is well suited.

The remainder of this chapter explores a framework to help you generate, evaluate and evolve valuable embedded projects. It uses the work of MIT's David Rose described in his book, *Enchanted Objects*. Rose's work helps to reveal potential blind spots for further investigation. It ensures habitual thinking (the result of our tacit privilege, prejudices and assumptions) is challenged. Even if one does not agree with details of Rose's approach, it is the process of reading and evaluating his work that is a useful tool to prompt analysis, reflection and imagination and avoid Sirius Cybernetics syndrome.

In the dedication of his book, Rose states that, "a more humane interface between technology and people is in your hands".

To identify what this means, he creates three lists of concepts that help one imagine, explore and assess ways in which embedded devices could be used. These lists are framed within the concept of "enchanted objects" that help one draw inspiration from magic, legends and fairy tales in which everyday things take on extraordinary properties. This re-framing of embedded development puts the focus on what the objects do, the capabilities they facilitate and how they solve people's problems. I particularly like how he emphasizes the contrast between the dynamic, liberating and imaginative outlook afforded by such thinking with the mass produced, impersonal and trite electrical gadgets so widely in use today.

The three lists address:

- human drives
- abilities of enchantment
- steps to enchantment

Be critical when evaluating these lists and try to work out how they apply to problems and opportunities in your users' lives.

Consideration of the lists will prompt insights for future projects using MicroPython and the devices described in the remainder of this book.

What will *you* build?

Human Drives

This list identifies six human drives that Rose believes are fundamental and universal: omniscience, telepathy, safekeeping, immortality, teleportation, and expression.[2]

Such drives are important because they provide clues to what makes a product, tool or device resonate with its users. They also categorize the sorts of broad problem domains you may wish to address, allowing you to learn from other efforts in the same context. As you read the descriptions of the items on this list, try to imagine how these drives manifest themselves in your life and consider how they are enhanced or confounded when you use an object, tool or device.

Omniscience
> *Omniscience* is knowing everything.

> People want to know about many different things from physics to the private lives of celebrities. Furthermore, people enjoy being the person who knows that special piece of valuable information because it gives them status. As Francis Bacon said, "knowledge is power".

> There are already objects that allow for the accumulation or delivery of knowledge. Objects, such as the barometer, forecast the weather and an ambient orb can be programmed to indicate stock market conditions through colourful displays of light.

> Rose mentions literary devices such as the alethiometer from Philip Pullman's trilogy, *His Dark Materials* (Yearling). This fictional object has the power to always reveal the truth.

> Is this fiction so far fetched?

> Try asking Amazon's Alexa, "what is the circumference of the sun in centimeters?"

> Perhaps it's not so far fetched after all.

> Ask yourself or your potential users to identify the nugget of information that's always needed at a glance? How would this information manifest itself in an object?

Telepathy
> *Telepathy* is communicating our thoughts, feelings and status to others.

---

2 Rose intentionally names concepts with a language that suggests folk lore, magic and the supernatural to prompt us to use our imaginations and consider "stuff" that isn't a manufactured opaque black plastic gizmo.

As the explosion of social media over the past decade has shown, people like sharing what they're up to, how they feel and their experiences and opinions with the rest of the internet. Some just crave attention.

Such interactions have the potential to widen or reinforce our social lives and promote a sense of community. It's how we collaborate with allies or encounter those who hold differing opinions. It allows us to broaden our view of the world.

Objects that make this possible include smart watches that connect to Twitter, mobile phones through which we have access to social media and voice communication and conference call devices built into meeting rooms. A fictional example is the Weasley clock from the Harry Potter books that keeps track of the status of members of the Weasley family (each member of the family is represented by a hand pointing towards a particular state, for example, sleeping, at work or lost).

Again, is this fiction so far fetched? Consider the many presence-based apps that use the GPS capabilities of your phone to advertise where you are and what you're up to.

Do you know someone who is absent but whose presence you desire? How might an object connect you? In what sense are you connected (for example, do you share your moods, locations or activities)?

Safekeeping

Safekeeping is the desire for protection from harm.

People thrive when they feel safe and at ease. It is a pleasant state and, if you are a parent or responsible for the care of others, it is an attribute you probably want to ensure for them.

In a digital sense, we expect the same feeling of safety.

We use devices to protect us from physical harm. For example, parking sensors warn us of impending contact with another vehicle. For our collective safety, smart motorways measure the traffic flow and change the speed limit to ensure a safe and speedy journey.

A literary example of an object that ensures safety is Frodo's sword "Sting" described in J.R.R. Tolkien's *Lord of the Rings*. It glows when there are Orcs in the vicinity.

Such benign devices need to collect data about us for them to work. But we also expect our data to be safe and for us to retain control over how it is shared.

Controversially, CCTV cameras watch our every move for what the authorities tell us is our own protection. One need look no farther than Orwell's *Nineteen Eighty-Four* to witness this phenomenon in literature.

Ask yourself, do you know anyone who needs help, care, and attention? What makes them require such care? Do you know of a vocation that is accident prone or requires health and safety guidance? What are the parameters for safety, and how might these be measured or highlighted by objects in the environment? What sort of data do these objects collect? How might they keep the data safe?

Immortality

Immortality pertains to the desire to lead a long and healthy life in which we can be autonomous, active, and attentive to the end.

This is perhaps an area where embedded devices are already well established through the technology used to promote and maintain health.

For example, do you use a Fitbit that tracks your daily activity to give you a "quantified self"? Perhaps you know someone with a pacemaker fitted to their heart. Rose provides the interesting example of a "glowcap", a device that fits to the top of medicine containers. It illuminates when the medicine should be taken, according to the doctor's directions; and, since it is connected to the internet, will reorder a prescription when supplies run low.

What aspects of your lifestyle do you want to enhance or diminish in order to lead a more flourishing life? Can you imagine an object to prompt, track, or discourage certain behaviours? Are there existing objects already associated with your health (or that perhaps cause ill health) that could be enhanced with behaviour from an embedded device?

Teleportation

Teleportation pertains to our desire for unconstrained movement.

We no longer live in a society where our world view is limited to our immediate geographical location. Transport usually involves complicated machines and procedures such as driving a car or catching a flight, so is it possible to make travel easier and perhaps less frustrating?

How could devices help us get to where we need to be?

A real-world innovation in this space are navigation handlebars for cyclists that glow on the left or right to indicate the direction to turn. Of course, within the next 10 years we're likely to see the emergence of self-driving cars, enchanted objects on wheels that know where you need to be.

Travel is already saturated with technical gizmos. How might such technology be simplified? What would you improve about a mode of transport, or perhaps even a transport hub, that made movement, arrival, or departure more bearable? How might the fabric of transport technology be changed from impersonal, grotty, and automatic machines to friendly, welcoming, and helpful objects that made travel a joy?

Expression

> *Expresion* is the need to make ourselves known through different artistic forms and media.

> Objects have always been a part of the creative drive, be they tools for building things, instruments for performing music, or perhaps the most well-known historic example of a revolutionary object for expressing yourself—the printing press.

> Today we have enchanted objects like 3D printers that will build anything, from minute engine parts to a whole house. Devices such as the Guitar Hero controller allow people to take part in musical activities that would otherwise require years of training. Lego Mindstorms, the Raspberry Pi, Arduino, and other platforms beloved by the maker movement allow people to create electronic and programmable projects that would have been impossible until very recently. They are examples of devices as a new programmable artistic medium.

> If you take part in an artistic endeavour, how might your tools or equipment be enhanced? Could you imagine a way in which a programmable object is a work of art?

Like me, you may be thinking that many of the examples associated with the human drives listed are already catered for by apps on the plastic and glass black mirror that is your smartphone.

I suspect Rose would counter by saying that bland, uniform applications on a mobile or tablet prevent access to some of the fun, excitement, and ease of use of real-world objects. The challenge is to figure out how such functionality can be migrated from a phone to an engaging, useful and valuable enchanted object. In doing so, the fabric of our world becomes more interesting, intriguing, and alive, rather than diminished into a device sitting in our pockets.

What would make such a world interesting, intriguing, and alive? Rose would answer with the abilities of enchanted objects.

Abilities of Enchantment

The seven abilities identified in this list collectively differentiate enchanted, embedded objects from other sorts of computing devices (smartphones, tablets, and PCs): glanceability, gestureability, affordability, wearability, instructibility, usability, and loveability.

Such attributes influence how we learn to use objects, how they interact with us, and how they inhabit the wider world. When reading this list, imagine how such abilities might manifest themselves in devices embedded in objects.

Fundamentally, these abilities relate to the experience of the user. By this I mean how attributes of an object affect a person's ability to use it to their advantage and how this, in turn, makes the user feel about the object. We're firmly in the world of human-computer interaction (HCI) or user experience (UX) for physical objects.

Glanceability

Glanceability helps us to read just enough information to make decisions.

This isn't a case of a pop-up, alarm, or other digital intrusion into our lives. Rather, it is bringing information into focus at the most opportune time and place.

Glanceability means less cognitive load and fewer interruptions, and quicker information-gathering and decision making. Such glances originate from us: we choose to look at a device rather than have it demand our attention through bleeps, vibrations, and visual interruptions.

Examples of glanceability include traffic lights and the hands on a clock face. People look at such devices when and where they need information. Such information is immediately conveyed via convention: stop at red or go on green.

Embedded devices that have lights, a display, or moving parts can be reconfigured so their physical properties or their appearance reflect or stand for some useful piece of simple information. To my British sensibility, objects that are glanceable feel more polite than attention-grabbing applications that compete for our attention.

Gestureability

Gestureability means an object senses and responds to our movements.

It may be that an object merely senses our proximity to engage some aspect of our environment, such as lights or heating. An object may be moved around in space to control some other object at a distance or respond to a gesture, such as a shake that represents a negative reaction. Perhaps an object reacts to rhythm in our movements or simply knows to enter quiet mode because it is face down.

All of this contrasts with typing on keyboards, poking and prodding touch-sensitive screens, or tediously dragging a mouse across a desk to point at visual metaphors such as a "button", "window", or "trash".

I wonder about the expressive potential of such capabilities. In the nondigital world, a violin translates and amplifies the movements of its player into music capable of profound emotional effect. It does this as a reaction to the highly practiced physical movements of the player's fingers and arms. Can you imagine a digital equivalent capable of similar expressive nuance?

Affordability
> *Affordability* makes embedded devices accessible to all.

Because the micro:bit is so cheap, it doesn't matter if a beginner programmer breaks it: affordability reduces risk.

Prototyping and development become less expensive, meaning more people can have a go. The falling cost of embedded devices, such as those upon which MicroPython runs, means there is more opportunity for bespoke, one-off projects whereby people create devices for their own particular niche.

Furthermore, the functionality required by most enchanted, embedded devices does not need the latest or fastest chips, video cards, or monitors. Such devices are within the price range of most people because their components are extraordinarily cheap.

Ultimately, affordability facilitates equality of opportunity.

Wearability
> *Wearability* is a way to liberate technology from beige boxes and black tablets to stuff that is around and even on us.

Much to the joy of its attendees, the Electromagnetic Field Camp (*https://www.emfcamp.org/*) (a community-organised maker conference in the UK) created conference badges that had an embedded display, WiFi, buttons, and the ability to run MicroPython. Never has a badge been so popular as attendees hacked together all sorts of previously unimagined uses for a conference badge.

At yet another recent UK-based Python conference, children eviscerated and then sewed back together cuddly toys, pendants, and micro:bits in order to make a programmable cyber-Teddy.

Many have taken microcontrollers and strips of NeoPixel LEDs to create costumes, hats, jewellery, and clothes that convey information (such as how "excited" the wearer is by measuring their heartbeat) or simply to look cool (such as the Knight Rider NeoPixel t-shirt[3]). Interactive clothing not only looks interesting, but can be useful as a new means of working with, placing, and controlling technology.

Indestructibility
> *Indestructibility* means embedded devices can last a long time.

This is in contrast to the latest $700 smartphone whose screen will often crack and which will be obsolete in terms of software within a couple of years.

---

3 Created by the talented Mr.Daniel Pope, for a fancy dress party. See: https://www.youtube.com/watch?v=UZr3oO5WXJI.

My local museum uses cheap embedded devices to run its push-button display cabinets, and they still work after a decade or more of constant use by the general public. By attaching an embedded device to a more robust everyday object, the electrical components have an extended lifetime.

Usability

Usability is how the shape, look, feel, and context of a device make its use self evident.

We all know how to sit in a chair, so could a chair become a user interface? If you have a bad back, why not embed a device that times how long you have been sitting? It could (glanceably) indicate when to stretch.

This is a classic example of how to repurpose everyday objects by enchanting them. In this way, technology draws upon the history, traditions, and familiarity associated with objects already found within our lives.

Loveability

Loveability is demonstrated by an emotional attachment to an enchanted object.

Perhaps the chair you modified to help manage your bad back originally belonged to Grandma, and, in turn, perhaps your modification will be cherished by your descendants as another aspect of the story of a family heirloom.

A device can also achieve loveability by looking friendly through anthropomorphic design or maybe because you simply appreciate its elegance in both form and function. In contrast, it's hard to become attached to a plastic and metal black mirror phone: while it is capable of doing many things, you know you'll get the same apps when you upgrade to the latest model.

The qualities described do not mean that objects containing embedded devices will be useful. It is quite possible to build something that suffers from the same inhuman, impersonal, and manufactured qualities that inhabit the many consumer electronics with which we are already familiar. As an alternative, Rose is encouraging us to step away from such mass-produced uniformity and, in its place, design embedded objects with personality, flair, and character. Thinking about the qualities enumerated will, he claims, help us make compelling stuff that resonates with people and intrigues users. Only then will devices become fun to use.

How should one go about addressing the human drives and enchanted qualities described? Rose has another list that enumerates the steps on the ladder to enchantment: a latter-day gradus ad Parnassum.[4]

---

4 Gradus ad Parnassum means "steps to Parnassus". Parnassus is a mountain in Greece that has a peak sacred to Apollo and the muses (the ancient Greek deities of the arts, sciences, and knowledge). As a title, it has historically been applied to guides in which one progresses to mastery through discrete steps.

Steps on the Ladder of Enchantment

What is the process of making an embedded object appear enchanted? Rose claims to have identified five repeatable steps on a "ladder of enchantment" that help bring forth objects and devices that meet the human drives and exhibit the enchanted qualities described. These steps are connection, personalization, socialization, gamification, and story-ification.

Their purpose is to act as a springboard for asking questions about how a device should function.

At the very least, Rose claims, climbing these steps will help you ask the *right sort of questions* as one progresses through an embedded project. Such steps are on a metaphorical ladder: the higher up the ladder one goes, the more sophisticated or enchanted the object is. He also makes it clear that not every object need reach the top step.

Connection

> *Connection* means adding sensing capabilities to an object.

> It may use such readings as triggers to signal information or store data away for later analysis. Such objects may also be connected to the internet and enhance their capabilities by offloading computation and data-storage to the remote computers "in the cloud".

> How could such sensing, processing, and connecting capabilities enhance the mundane tools or devices we use in everyday situations? What important information do we require? How is this to be gathered? What happens to the collected data?

Personalization

> *Personalization* means reacting to context.

> Given the collection of data in the previous step, how could this be used to make the device work better? Could an object learn from and about its context or environment so that it modifies and improves its behaviour? What needs to be measured? What aspect of the device should change? How should the change relate to the detected context?

Socialization

> *Socialization* means adding connections to wider society: friends, family, and colleagues.

> It may also mean connecting objects to make a useful or valuable function an ensemble effort. For example, the ringing of an enchanted dinner gong may involve reaching out to your family and the devices on or near them.

What do you wish to signal and to whom? How are objects used to trigger such signals? How do you indicate recipients? Is it possible to modify the signal with extra information? How is the signal brought to the attention of recipients? How do you ensure the privacy of such communication?

Gamification

Gamification means adding fun, motivation, and the occasional nudge to the behaviour of a device.

The aim is to help people become active users and participants in the full capabilities of an object. This could be achieved with tropes found in video games (a scoring system, achievements, and so on), although sensitivity and respect should be involved to ensure the device does not behave inappropriately. Think of the enforced "help" that Clippy attempted to bring to Microsoft Word.[5]

Is the object supposed to be fun to use? If so, how? If not, why not? What aspects of the device need effort from the user? Does this require you to nudge users to activate and learn such functionality? When and how should you nudge? What does the object do to help overcome potential user frustrations?

Story-ification

Story-ification is adding a narrative that sets the object in a meaningful context.

In a sense, objects could become characters in the stories of their users' lives by providing meaningful interactions. Alternatively, the object itself may come with a back story that makes it interesting and unique.

To understand how powerful a story can be, just look at how children weave objects into their play: that stack of Lego bricks is really a rocket ready to take them to Mars, or the stick found in the park is a magic wand that only works on grandparents.

This also applies to how objects relate to important narratives in the real world. Only people with an employee card get into the office (excluding people who don't have an object), "with this ring, I thee wed" (binding people together through an object), and only police officers carry a badge (an object that establishes status). Such stories about objects give our world meaning.

---

5 Clippy was a digital assistant built into versions of Microsoft Office in the late 1990s. It took the form of a panglossian paper clip. At inopportune moments Clippy would pop up to cheerfully interrupt your workflow with helpful suggestions like, "It looks like you're trying to write a letter, would you like me to..." (followed by a list of obvious or unhelpful outcomes). Like the doors with built-in personalities from *The Hitchhiker's Guide to the Galaxy*, Clippy was universally mocked and loathed.

Does the enchanted object exist within a narrative? How is this exhibited in the way the object behaves? Is it obvious what the story is? How do people learn or make up the story related to the device?

David Rose's stated aim was to give people a way to develop fresh ideas for creating embedded devices that engaged people in a more humane way than terminals, GUIs, or touchscreens. Such enchanted objects that contain relatively simple embedded devices are not an end in themselves. Rather, their usefulness comes from their ability to sense, adapt, connect, motivate, and tell a story as an object that solves a problem, creates value, provides opportunities, or empowers its users.

What do you need to know about embedded devices and MicroPython to create such things? That is what the rest of the book will explore.

## CHAPTER 7

Visual Feedback

One of the most deeply rewarding aspects of programming is making the computer appear to do something; be it blinking an LED, printing "Hello, World!", displaying a picture, or creating an animation, there's something satisfying in making such output visible to the world.

Blinkenlights

Blinkenlights[1] are the embedded world's version of `"Hello, World!"`. If it's possible to make an onboard LED blink on and off, then you have probably got everything set up correctly (as demonstrated in the earlier chapters of this book).

This activity might at first appear boring, since an LED's purpose is either to be decorative or indicative of state. Crucially, familiarity with such an innocuous component provides an interesting route into MicroPython's modus operandi. It demonstrates how MicroPython is both similar and very different to "normal" Python. This journey of discovery starts with the LED itself.

Light emitting diodes (LEDs) come in many colours and all work in the same way. From the perspective of physics, an electrical current of a suitable voltage is applied causing the occurrence of electroluminescence. In the case of an LED, a semiconductor material emits light in response to the electrical current. The colour of light is

---

1 Fake-German for blinking lights, such as LEDs, that may indicate some sort of status. The original term first appeared as a mangled set of mock-German instructions found on a 1950s-era sign in an IBM computer lab. The sign ended with, "RELAXEN UND WATSCHEN DER BLINKENLICHTEN". Interestingly, the German Chaos Computer Club created project blinkenlights (*http://blinkenlights.net/*), which turned the windows of a high-rise block in central Berlin into a monochrome display in which the blinking lights, one in each window of the building, displayed animations and messages.

determined by the energy of the released photons. This, in turn, is determined by properties of the electroluminsecent semiconductor. Different semiconductor materials have different properties and thus produce different colours. The amount of energy applied to the LED changes its brightness.

The microcontrollers upon which MicroPython runs have pins that can accept input from or provide output to external components in the form of a voltage, which in turn provides an electrical current. The LEDs built into a device, or the LEDs connected via externally available GPIO pins, are managed by MicroPython controlling the flow of electrical current to the pins attached to the LED.

MicroPython runs on the "bare metal". This means MicroPython is the broker between our Pythonic view of the world and the hardware connected to the microprocessor.

As your Python code executes, MicroPython directly controls the physical aspects of the device by applying or detecting a voltage (and hence electrical current) on the microcontroller's pins.

Microcontrollers are relatively simple and don't need to run a complicated operating system kernel like Linux. There is no operating system because MicroPython *is the operating system*. The pins and hardware-related protocols used to run peripherals attached to the microcontroller are available as Python-based APIs (such as the `pyb`, `microbit`, `digitalio`, and `machine` modules used to blink the LEDs in the previous chapters and to be explored throughout the rest of this book).

To give a flavour of how this relates to our humble LED, let us examine how it is possible to change the amount of power sent to an LED, thus adjusting its brightness.

Since programming means working in the digital world of "on" and "off", there is no notion of "in between". For example, we cannot turn on the LED at only 50% of full brightness. Instead, we change the brightness of an LED using a technique called *pulse width modulation* (PWM). This important concept is one way to make something appear analog (where there is a graduation of possible values, such as varying degrees of brightness) when the underlying implementation is digital (there are only two possible values: "on" and "off"). This is achieved by switching the voltage of a pin on and off very quickly whilst also controlling the length of time that the voltage is on or off.

Figure 7-1 demonstrates what I mean.

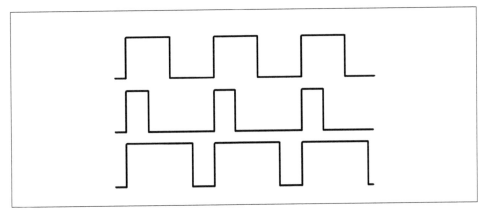

Figure 7-1. Three different PWM signals.

The three signals have the same period (frequency), but they have different *duty cycles*, the amount of time in one period when the signal is on. Because power is only applied when the duty cycle is on, the average power delivered to the device is the *duty cycle ratio* multiplied by the "on" power.

For example, imagine the signals were "on" at 3.3 volts and "off" at zero volts. The first signal is on only 50% of the time, so the total energy of the signal is half that of a fully "on" signal. This has the equivalent effect of outputting an average of 1.65 volts instead of 3.3 volts. The second signal has a 25% duty cycle, as if an average of 0.825 volts were output. The third signal has a 75% duty cycle, three times as much energy as the second signal and therefore equivalent to an average output of 2.475 volts.

Peripherals that benefit from PWM tend to have a slow reaction time to the input signal. Because of their slow reaction time, they appear to operate smoothly, even though power is applied to them in pulses rather than continuously.

Unfortunately, LEDs are fast reacting. The LED blinks as a result of power applied during the duty cycle of the PWM signal. However, the human eye is too slow to notice that the LED is blinking. The LED simply appears constantly illuminated because it blinks faster than the human eye can detect. Since the average voltage of a PWM signal relates to the amount of light emitted, the brightness of the LED appears to change.

Therefore, the smoothing out of pulses into a seemingly continuous analog signal is the result of either a relatively slow-reacting peripheral that can't display a discernible reaction to individual pulses, or a slow-reacting observer of a fast-reacting peripheral, such as an eye observing an LED.

As we shall discover in later chapters, PWM works well for controlling motors and can be used to generate sound waves.

Not all devices or LEDs work with PWM. The following example, using the REPL on the PyBoard, illustrates the simple on/off activity of the device's red LED:

```
>>> red = pyb.LED(1)
>>> red.on()
>>> red.off()
>>> red.toggle()
```

The red LED on the PyBoard is labelled as 1, with the green labelled 2, amber as 3, and blue as 4. The red and green LEDs can only be on or off. However, the amber and blue LEDs are capable of using PWM to change the intensity of their brightness, like this:

```
>>> blue = pyb.LED(4)
>>> i = 0
>>> while True:
...     pyb.delay(5)
...     i += 1
...     if i > 255:
...         i = 0
...     blue.intensity(i)
...
```

Notice how the pyb.delay function is used to pause the script for 5 milliseconds so the gradual changes in brightness do not occur too quickly for the human eye to see them fluctuate. Valid intensity levels (represented by the object i) are between 0 (full off) and 255 (full on). Once the intensity of brightness is higher than the maximum value, it is reset to 0.

On the micro:bit, brightness is simplified to only a few levels of possible intensity. Each of the LEDs in the display matrix can be set to a value between 0 (off) to 9 (brightest).

The brightness of the red LED on the Circuit Playground Express is also controllable via PWM. The API to do this in CircuitPython makes explicit reference to PWM terminology. The following example illustrates what I mean:

```
>>> from board import D13
>>> import time
>>> import pulseio
>>> pin = pulseio.PWMOut(D13)
>>> while True:
...     for i in range(16):
...         pin.duty_cycle = 2 ** i
...         time.sleep(0.1)
...
```

The pin to which the LED is connected is represented by the D13 object imported from the board module. Python's standard time module is imported for later use to introduce a pause in the program so that the gradual changes in brightness don't happen too quickly. However, the `pulseio` module is most interesting because it contains the `PWMOut` class. By instantiating the class with the reference to the pin, a `pin` object is created. The `duty_cycle` attribute (note the correct use of terminology here!) can be any value between 0 and 65,536 (i.e., a 16-bit resolution). As the value of the duty cycle is increased, so is the LED's brightness.

Unfortunately, the LEDs on the ESP8266/32 are digital (on/off) only.

While the Circuit Playground Express has an LED that can blink, it also has a secret weapon when it comes to lights.

NeoPixels (Blinkenlights on Steroids)

NeoPixels are multicolour LEDs.

Whereas the LEDs discussed in the previous section only display light in a specific and unchanging colour, NeoPixels can emit more than 16 million different colours.

All the boards discussed in this book are capable of working with NeoPixels, but only Adafruit's Circuit Playground Express has them built in. For the other boards, you will need to purchase rings, strips, boards, and sticks of NeoPixels and connect them to the GPIO pins as per the peripheral's instructions. In any case, the `neopixel` module you will use in MicroPython works in roughly the same way on each device.

Each NeoPixel is actually three LEDs and a driver chip combined into a very small form factor. Each of the constituent LEDs emits only one colour of light, either red, green, or blue. By controlling the intensity of the individually coloured constituent LEDs, it is possible to mix the colours into one of over 16 million possible combinations (see Figure 7-2).

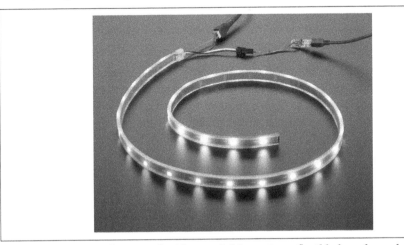

Figure 7-2. A strip of NeoPixels. Also available as rings, flexible boards, and sticks.

The results are spectacular and give most Christmas trees a good run for their money.

NeoPixels are also fun when you think about the signalling potential of the Circuit Playground device or the various rings, sticks, and boards of NeoPixels that could be attached to the other devices. If something has gone wrong, flash red for danger; if it's getting cold, turn blue; if something's confusing, blink lots of different colours. The possibilities are only bounded by your imagination, and NeoPixels certainly inspire creative use of light.

The neopixel module (found in CircuitPython's bundle of useful libraries, mentioned earlier) provides a single class called NeoPixel. An instance of this class is used to drive the physical NeoPixels. In order to instantiate the NeoPixel class, you will need a reference to the pin object representing the pin to which the DIN connection to the NeoPixels is connected along with an indication of the number of NeoPixels connected via the referenced pin. On the Circuit Playground Express, the code looks like this:

```
import neopixel
from board import NEOPIXEL

np = neopixel.NeoPixel(NEOPIXEL, 10)
```

In this fragment, the np object is created from the NeoPixel class. A reference to the pin for the NeoPixels is found in the board module[2] (it's helpfully called NEOPIXEL) and we already know there are 10 NeoPixels available for us to use on the Circuit Playground Express, hence the value of the second argument.

This demonstrates one of the most common ways in which MicroPython allows you to interact with attached hardware: instantiate a class representing the high-level functionality of the attached peripheral. Methods on the resulting object abstract away the low-level implementation details and allow you to concentrate on the useful behaviour of the device. Low-level attributes are set at instantiation, such as references to the pins to which the peripheral is connected.

In the preceding example, the resulting np object is an addressable container: it acts like a list and allows you to get and set individual pixels using the square brackets of subscription notation.[3] In this way it's possible to reference individual NeoPixels by position and get or set their associated red, green, and blue (RGB) values, as the following REPL session (based on the earlier code fragment) demonstrates:

---

2 References to all the pins on a CircuitPython device are always found in the board module. We cover how to reference pins on all the devices covered in this book in Chapter 9 (although you'll see examples of pin references prior to that chapter). For example, on a micro:bit you may use pin0 found in the microbit module.

3 Under the hood, this is implemented by the __getitem__ and __setitem__ special methods.

```
>>> import neopixel
>>> from board import NEOPIXEL
>>> np = neopixel.NeoPixel(NEOPIXEL, 10, auto_write=False)
>>> np[0]
(0, 0, 0)
>>> np[0] = (32, 0, 32)
```

The colour of a NeoPixel is represented by a tuple containing three integers that must be in the range of 0 to 255. They represent the intensity of the RGB light emitted by the referenced NeoPixel. In the previous example, immediately after instantiation of the np object, we read that the first NeoPixel (in position 0) is not emitting any light, represented by the tuple (0, 0, 0). The next line updates the value associated with the same NeoPixel to a tuple where the intensity of red and blue light is set to 32. (The order of the numbers in the tuple is always red, green, blue.)

If you are following along with your own device, you're probably wondering why you can't see a purple glow from the NeoPixel in position zero. It's because we instantiated the NeoPixel class with the auto_write flag as False. The Adafruit version of the neopixel module automatically updates the state of the physical neopixels as you make changes in code. All other implementations of the neopixel module require you to explicitly call the write or show method for your changes to take effect on the hardware. For the sake of consistency between implementations, I'm making the Adafruit version of the module behave like the others in the following code examples.

```
>>> np.write()
```

 The write method was renamed to show in the micro:bit implementation of the neopixel module, since it was felt this would make it easier for children to understand what was going on. The write method is an alias of show in the Adafruit version of the module for exactly the same reasons whilst also retaining backwards compatibility.

If you need to act upon all the NeoPixels at the same time, there is a convenience method called fill that takes a single tuple and applies it to all the NeoPixels at once:[4]

```
>>> np.fill((0, 32, 32))  # Turn all NeoPixels to cyan
>>> np.write()
```

This is also useful if you need to switch off all the NeoPixels:

```
>>> np.fill((0, 0, 0))
>>> np.write()
```

---

4 This isn't supported in the micro:bit version of the neopixel module.

Given such a simple yet spectacularly colourful API, it's fun to combine NeoPixels with a program loop so the device appears alive:

```
import neopixel
import random
import time
from board import NEOPIXEL

np = neopixel.NeoPixel(NEOPIXEL, 10, auto_write=False)
step = 32

while True:
    for i in range(10):
        for j in range(10):
            np[j] = tuple((max(0, val - step) for val in np[j]))
        r = random.randint(0, 255)
        g = random.randint(0, 255)
        b = random.randint(0, 255)
        np[i] = (r, g, b)
        np.write()
        time.sleep(0.05)
```

This fragment cycles around the available NeoPixels and, for the next available Neo-Pixel, selects a random colour whilst dimming the intensity of any active NeoPixels by a step of 32. The effect is a randomly coloured circular disco light. Change the value passed to time.sleep to change the speed of the effect. Alternatively, change the value of step to control how quickly the NeoPixels dim (this will also influence the length of the "tail" of the circular movement around the display).

It demonstrates how easy it is to create colourfully decorative or informative devices (flash red for danger!).

Another way to visually communicate information is through words and pictures, and two of our target devices make this very easy.

Text, Images, and Animation

The micro:bit has a limited but remarkably flexible display consisting of a 5 x 5 matrix of red LEDs. As mentioned earlier, these LEDs can be in one of 10 possible settings numbered from 0 (no light) to 9 (brightest). The lower-level PWM used to change the brightness of the LEDs is hidden from the user: they need only concern themselves with an abstract notion of brightness levels.

As noted in Chapter 3, the microbit.display object is used to set (and get) the value of an individual LED "pixel". However, the display object, in combination with the microbit.Image class, allows you to create all sorts of interesting effects.

The microbit.display object is also an example of yet another way in which Micro-Python exposes hardware: an object is created to represent an unchanging aspect of the hardware. For example, every micro:bit will always have an LED matrix. Rather than instantiating a class each time you need to work with the display, as a convenience, MicroPython provides a useful object to represent the display. All interactions with the display therefore happen via methods belonging to the built-in `micro bit.display` object.

Perhaps the most immediately interesting feature is scrolling textual output:

```
from microbit import display, Image

display.scroll("Hello, World!")
```

You get plenty of options with the `display.scroll` method: it's possible to change all sorts of aspects of the output. Pass the `delay` parameter to control the number of milliseconds between each update of the display. This controls how fast the text scrolls. The boolean `wait` parameter (whose default value is `True`) indicates if the code should block until the scrolling has finished. If the `loop` parameter is `True`, scrolling will repeat forever. Finally, if the `monospace` parameter is `True`, the characters will take up 5 pixel-columns in width; otherwise, there will be exactly one blank pixel column between each character as they scroll.

For example, the following fragment[5] demonstrates how to make a monospaced message continually scroll quickly while the device is able to get on with other things:

```
display.scroll("Hello!", delay=80, wait=False, loop=True, monospace=True)
```

If you want the characters to display one after the other, rather than scrolling, use the `show` method instead:

```
display.show("Danger!")
```

Apart from the inapplicable `monospace` parameter, the `show` method understands all the same parameters as the `scroll` method. Furthermore, it understands the `clear` parameter, which if `True`, clears the display after the animation has completed.

But it's not just characters that can be displayed. The micro:bit has an `Image` class that makes it very simple to create pictures and make use of a large number of pre-built images (see Figure 7-3). For example, here's how to display a happy face:

```
display.show(Image.HAPPY)
```

---

5 From now on I assume you've included the `from microbit import *` (or equivalent) line.

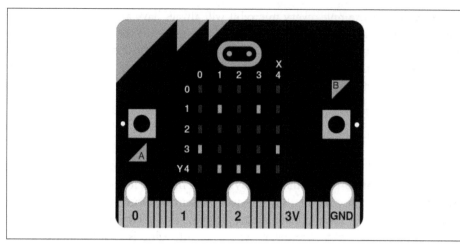

Figure 7-3. A happy looking micro:bit.

There are 65 built-in images: the uppercase attributes of the Image class. They are all listed in the documentation (*https://microbit-micropython.readthedocs.io/en/latest/ image.html*), although you could discover them all from the REPL by using the dir(Image) command. How do we know these pictures appeal to a young beginner coder? Because they were mostly designed by an 11-year-old girl based in the UK!

But you may want to create your own work of art. For this, you will need to create an instance of the Image class. Here's an example of how to create an interesting pattern with some LEDs brighter than others:

```
from microbit import display, Image

my_picture = Image(
    '33333:'
    '36663:'
    '36963:'
    '36663:'
    '33333:')

display.show(my_picture)
```

The results will look something like Figure 7-4.

When instantiating the Image class, pass in a string containing numbers representing the brightness of each LED "pixel" starting from the top left and ending at the bottom right. A colon (:) is used to represent the end of a line. The way I have formatted the code in the preceding example allows you to see how this relates to the LED matrix on the front of the micro:bit. You could just as easily have written the string as '33333:36663:36963:36663:33333:'.

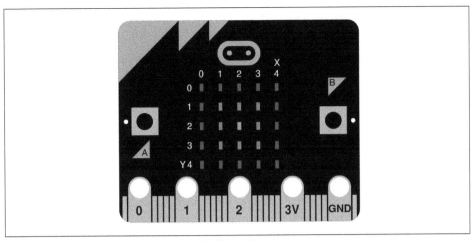

Figure 7-4. A micro:bit with pixels of different brightness.

An instance of the `Image` class need not be bound by the dimensions of the display. This may be useful if, for example, you are creating some sort of maze-like game where only a subsection of the maze is to be displayed at any one time on the micro:bit's display. In this case, you could use a very long string, in a manner similar to the one demonstrated, but covering an image of much larger dimensions. Alternatively, you could use the second form of creating an `Image` object:

```
from microbit import display, Image

buf = bytearray(x % 10 for x in range(100))
i = Image(10, 10, buf)
```

Using Python's built-in `bytearray` type (a way to very efficiently store bytes), we store numbers between 0 and 9 in a buffer to indicate the intensity of LEDs in an image. Combined with an indication of the width and height, it's possible to create a new image. In the previous code example, it's a 10 x 10 image where each row contains pixels that get brighter the further they are to the right.

Given such an image, it is possible to slice and dice it in interesting ways to generate new images. For example, to create a 5 x 5 viewport from a specific location in the larger image (in order to show it on the display), use the `crop` method to return a new (temporary) image to display:

```
display.show(i.crop(3, 4, 5, 5))
```

The `crop` method needs the X and Y coordinates of the top-lefthand position of the new image within the original image, along with a width and height. In the example above, the X (3) and Y (4) coordinates come first followed by the width and height (both 5). If you were creating a game with a maze larger than the screen, you would use `crop` to shift the display viewport over the image, representing the maze as the player moves around (in some yet-to-be determined manner).

Instances of the `Image` class come with other useful methods such as, `copy` (to return an exact copy of the image), `invert` (to return a new image by inverting the intensity of the pixels in the source image) and `fill(value)` (to set all the pixels to the integer `value`). Furthermore, once an `Image` instance is created it is possible to modify and read specific pixel values as this REPL-based session demonstrates:

```
>>> i.get_pixel(1, 2)
1
>>> i.set_pixel(1, 2, 9)
```

The common pattern is the specification of the X and Y coordinates of the referenced pixel in the image. Just pass these to return the current value of a specific pixel via the `get_pixel` method. Additionally, supply a new valid value when you use `set_pixel`. Note, updating the underlying image will not update the display—you will need to refresh it with a suitable call to `display.show` the modified image.

If you have an image that is too wide or high it is possible to use various `shift_[up, down, left, right]` methods to make scrolling very simple to achieve by supplying an integer indicating the offset for the new image. Internally, this is how characters are scrolled with the `display.scroll` method.

Finally, all the shift and `crop` methods are built upon a single, but very powerful, method called `blit`. It takes a source image and a definition of a rectangle. It updates the `Image` instance with the light intensity values defined in the area covered by the specified rectangle in the original image. For example, the `crop` method could be reimplemented like this:

```
def crop(self, x, y, w, h):
    result = Image(w, h)  # Create a new empty image called "result".
    result.blit(self, x, y, w, h)  # Blit from self into result.
    return result  # Return the new image containing the crop.
```

The remaining piece of the micro:bit display puzzle is animation and, because of the educational heritage of the device, it is very simple.

Use an iterator object (an object that is able to keep giving new items) that returns instances of the `Image` class. These instances will act as frames in your animation. The iterator object—the source of frames—is passed into the `display.show` method used to show pictures. For example, a list of `Image` instances can act as a source of animation frames. The `Image` class has two lists of such images already available: `ALL_CLOCKS` (containing representations of all 12 positions of a clock hand) and `ALL_ARROWS` (containing arrows pointing to the most common headings of a compass). To animate the hand of a clock on the micro:bit display, do this:

```
display.show(Image.ALL_CLOCKS, delay=50, wait=False, loop=True)
```

The effect is something akin to a radar display from a cheap 1970s sci-fi TV show.

You could, of course, use a generator function to keep creating new Image-based "frames" in the animation:

```
from microbit import *
import random
import array


def animation():
    blinkenlights = array.array('b', [random.randint(0, 9) for i in range(25)])
    yield Image(5, 5, blinkenlights)


while True:
    display.show(animation())
```

The result of which is a blinkenlight display like the computer "WOPR" found in the 1983 movie *War Games*.

The simplicity and limited capabilities of LED displays are evident when compared to the full-colour displays we are used to seeing in our phones, tablets, and laptops. The sorts of devices that require such display capabilities do not run MicroPython, but it doesn't mean MicroPython can't drive a relatively simple version of such displays. The PyBoard's LCD display skin is a great example of this capability in action.

PyBoard Colour LCD Display

The colour LCD display (lcd160cr) is an interesting peripheral for the PyBoard. Not only is it a colour display, but it's also touch-sensitive, so it can act as both an input and output device—a sort of micro-version of an iPad (iPad nano?).

It is possible to connect the LCD display in two positions: X and Y (as shown in Figure 7-5).

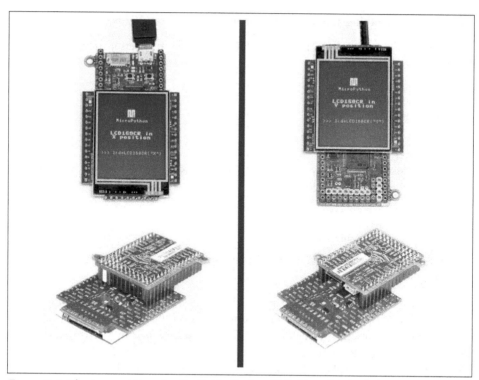

Figure 7-5. The two positions for the LCD display: X (left) and Y (right).

Once connected, the simplest way to start using the peripheral is to run the demo from the REPL:

```
>>> import lcd160cr_test
To run all tests: test_all(<lcd>)
Individual tests are: test_features, test_mandel
<lcd> argument should be a connection, eg "X", or an LCD160CR object
>>> lcd160cr_test.test_all('X')
```

As the instructions explain, call the `test_all` function with the appropriate position (X or Y). You will see a few seconds of a graphical readout of the display's features followed by a blue Mandelbrot set. If these appear, you are ready to go. If not, sometimes it's good to switch things off, unplug the display, and plug it in again firmly and restart.

It is very simple to get things onto the screen:

```
import lcd160cr


lcd = lcd160cr.LCD160CR('X')
lcd.erase()
lcd.write('Hello, World!')
```

An object representing the physical LCD (lcd) is instantiated with the LCD160CR class and an indication of how the device is connected ("X" or "Y"). Once instantiated, the physical device will come to life and display the MicroPython logo to indicate that it is working. Often you don't need this, so the erase method clears the screen. Pass in a string to print text on the screen via the write method. The end result is the message rendered as small green writing in the top-lefthand corner. We have just created the world's smallest retro-monochrome, green-screen monitor! How 1337 is that?[6]

Not as 1337 as piping the REPL to the display:

```
>>> import lcd160cr
>>> import pyb
>>> lcd = lcd160cr.LCD160CR('X')
>>> lcd.erase()
>>> uart = pyb.UART('XA', 115200)
>>> pyb.repl_uart(uart)
```

At this point, all interactions with the REPL appear on the display. You too can bask in the glory of looking like a miniature version of a Hollywood hacker.

More seriously, given the small form factor and even smaller text size, we should try to make things readable by changing the font. There are four possible fonts, with the default output being the smallest (0). Since you are probably already in the REPL piped to the display, try changing the font to a value between 0 and 3 (inclusive):

```
>>> lcd.set_font(1)
```

The set_font method accepts optional arguments to change the look of the textual output, all of which have a default value of 0. The scale argument defines the scaling of the pixels used to render the text. If a pixel is square, its side is equal to scale + 1, with scale's maximum valid value as 63. The bold argument can be a value between 0 and 31; and as the value increases, so does the effect of boldness on the text. The trans argument is a flag to determine if the characters are rendered with a transparent background (1) or not (0). Finally, scroll is another flag to set if the display should do a soft scroll (i.e., the text gradually moves up the screen) if the value is set to 1, or a hard scroll (the text immediately moves up the screen) if the value is set to 0. Continuing the REPL-based example, we can set the font to font-family 3 with no scaling, a slightly bold effect, zero transparency, and a cool smooth scrolling effect:

```
>>> lcd.set_font(3, scale=0, bold=1, trans=0, scroll=1)
>>> print("Hello, World!")
```

Use the set_text_color method to change both the text and background colours. The colours need to be expressed as a 16-bit integer that represents the red, green,

---

6 "Elite" written using "leet" spelling. See *https://en.wikipedia.org/wiki/Leet*.

and blue values. Happily, the `LCD160CR` class has a static method called `rgb` to do this for you:

```
>>> text_colour = lcd.rgb(255, 128, 0)
>>> background = lcd.rgb(0, 128, 255)
>>> lcd.set_text_color(text_colour, background)
>>> lcd.write("Hello, World!")
```

Finally, to override the position of text, you need to provide the X and Y coordinates for the upper-left corner of the new block of text:

```
>>> lcd.set_pos(20, 40)
>>> lcd.write("Hello, World!")
```

While textual output is important, it is graphical output that turns heads.

At a pixel-related level, the API is remarkably similar to that of the micro:bit. It's possible to `set_pixel` and `get_pixel` given X and Y coordinates and, when setting the colour, an RGB value generated in the same way as for textual colour. The following script demonstrates this sort of capability with a multicolour "snow crash"[7] effect:

```
import lcd160cr
import random

lcd = lcd160cr.LCD160CR('X')
lcd.erase()

while True:
    r = random.randint(0, 255)
    g = random.randint(0, 255)
    b = random.randint(0, 255)
    colour = lcd.rgb(r, g, b)
    x = random.randint(0, lcd.w)
    y = random.randint(0, lcd.h)
    lcd.set_pixel(x, y, colour)
```

Notice the use of the `lcd.w` and `lcd.h` constants that define the display's pixel width and height, respectively.

There are also methods for drawing simple shapes and lines. These rely on the important concept of the *pen*. The pen has both a line and fill colour. The *line colour* defines the outline colour of a shape, whereas the *fill* is the colour inside the shape. Drawing a shape is as simple as defining the pen's attributes with `set_pen` and drawing with one of the two sorts of primitives: rectangles or lines. This continuation of the REPL session demonstrates the basic methods you need:[8]

---

7 A "snow crash" is the static noise you get on the screen of an old mis-tuned analog television.

8 There are a large number of methods attached to the `LCD160CR` class that use the pen. They are described fully in the MicroPython documentation (*https://docs.micropython.org/en/latest/pyboard/library/lcd160cr.html*).

```
>>> lcd.reset()
>>> lcd.set_pen(lcd.rgb(255, 0, 0), lcd.rgb(0, 0, 255))
>>> lcd.rect(20, 20, 40, 40)
```

In the preceding code example, the pen's outline colour is set to red and the fill colour is set to blue. An equilateral rectangle[9] is drawn onto the display: its top-left corner is defined by X and Y coordinates (20, 20) and width and height lengths (40, 40).

Given such simple primitive operations, it is very easy to quickly create something with lots of visual appeal. For example, here's a very simple script to continuously generate pictures in the style of the famous Dutch artist, Piet Mondrian:

```python
import pyb
import lcd160cr
from random import randint, choice, uniform

lcd = lcd160cr.LCD160CR('X')

MAX_DEPTH = 4
RED = lcd.rgb(255, 0, 0)
YELLOW = lcd.rgb(255, 255, 0)
BLUE = lcd.rgb(0, 0, 255)
WHITE = lcd.rgb(255, 255, 255)
BLACK = lcd.rgb(0, 0, 0)
COLOURS = [RED, YELLOW, BLUE, WHITE, WHITE, WHITE]


class Node:
    """
    A node in a tree representation of a Mondrian painting.
    """

    def __init__(self, depth=0):
        """
        Choose the colour of the rectangle, work out the depth
        add child nodes if not too deep.
        """
        self.colour = choice(COLOURS)
        self.depth = depth + 1
        self.children = []
        if self.depth <= MAX_DEPTH:
            self.direction = choice(['h', 'v'])
            self.divide = uniform(0.1, 0.9)
            self.children.append(Node(self.depth))
            self.children.append(Node(self.depth))

    def draw(self, x, y, w, h):
        """
```

---

9 Also known as a square.

```
Recursively draw this node and its children.
"""
lcd.set_pen(BLACK, self.colour)
lcd.rect(x, y, w, h)
if self.children:
    if self.direction == 'h':
        self.children[0].draw(x, y, int(w * self.divide), h)
        self.children[1].draw(x + int(w * self.divide), y,
                              int(w * (1.0 - self.divide)), h)
    else:
        self.children[0].draw(x, y, w, int(h * self.divide))
        self.children[1].draw(x, y + int(h * self.divide), w,
                              int(h * (1.0 - self.divide)))

while True:
    # Keep re-drawing new Mondrian pictures every few seconds.
    tree = Node()
    tree.draw(0, 0, lcd.w, lcd.h)
    pyb.delay(randint(4000, 8000))
```

The results will look something like Figure 7-6.

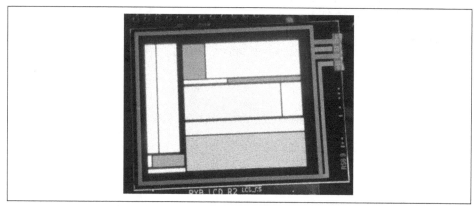

Figure 7-6. MicroPython does Mondrian

I hope you agree, the picture looks quite convincing.

Such visual output is immediately striking for the user. An equally striking yet contrasting capability is to program a device to sense its environment. As a result, it comes alive by reacting to things external to itself.

## CHAPTER 8

Input and Sensing

It is important for a device to react to interactions from users or changes in its environment. Such interactions are at the core of the device's nonprogrammer user experience. For example, shaking the device to reset the state of the program, using the light sensor to change the intensity of the LEDs, or pressing a button to cycle through a series of options are all interactions that rely on the device processing input or sensing changes in its environment.

The devices described in this book do not have keyboards, mice, or touchscreens like traditional computing devices. So how are we to interact with them? Typing Python commands into a REPL might be fun for programmers, but this doesn't address the needs of nontechnical users who need to communicate intents, instructions, or decisions in a manner that is intuitive, context sensitive, and perhaps even entertaining. Therefore, it's important to consider how MicroPython works with input and sensors.

Collectively, the devices covered in this book have a wide variety of input and sensing capabilities available to them. Some, like the ESP8266/32-based devices, are limited only to one or two buttons; others, like the Circuit Playground Express and micro:bit, have all sorts of components through which users can interact with the device. In all cases, it is possible to connect external sensors and peripherals to the boards via GPIO pins, and the general prinicples of working with such hardware can be transferred between devices.

We start our exploration of input and sensing with the ubiquitous button.

Buttons and Capacitative Touch

Buttons are interesting in an, "Ooh, I wonder what happens if I press this?" sort of a way. As programmers, it's our job to make sure people are not disappointed with the result of pressing a button!

Every device has at least one reset button that will restart the board. Some devices have several more buttons.

The micro:bit has the simplest means of interacting with its two buttons (labelled A and B). These are represented by two objects found in the microbit module called button_a and button_b. Both button objects have the same methods, and in order for them to be useful, we need to consider the notion of an event loop and containing a short pause.

A fundamental requirement when dealing with input is waiting for something to happen. This is achieved with an *event loop* (code that continuously loops around waiting for and handling input events). Furthermore, when user-generated events occur, they often need to be handled in a way that works in human, rather than computer, time frames. For example, when you press a button, the event loop will have cycled lots of times in the fraction of a second you will have taken to actually press the button. What happens if we only want the button-press to be handled once, rather than on each of the numerous iterations of the event loop that occured while we pressed the button? The solution is to insert a short pause to the event loop to slow it down so multiple events don't fire quickly when we only need to register a single event.

The next example illustrates both concepts:

```python
from microbit import *

position = 2
while True:  # event loop
    sleep(60)  # pause
    if button_a.is_pressed():
        display.clear()
        position = max(0, position - 1)
    elif button_b.is_pressed():
        display.clear()
        position = min(4, position + 1)
    display.set_pixel(position, 2, 9)
```

In order to wait for something to happen, we simply make an infinite loop around a piece of code that defines how to react to certain expected events (such as a button press). In the example, the infinite loop is achieved in the simplest possible manner with while True:. The resulting blocks of code are conditional on button presses. If a button is pressed, the code changes the position value and then displays a pixel in that position in the X-column and in Y-row three. It's a very simple means of moving the pixel from left to right. Perhaps the best way to understand what the "pause" line is for is to remove it and try to use the buttons.

The problem is the event loop is way too quick for our human reflexes. If there were no pause line to slow down the cycle of the loop, then you would only be able to move the pixel to the extreme edges. Why? Because your reaction for pressing a button is slow enough that the event loop will cycle too many times (thus moving the pixel to the extreme left or extreme right). The pause gives you just enough time to tap the button to move the pixel by a single unit. An interesting way to explore this feature is to change the number of milliseconds the device sleeps as part of the pause operation and observe how this affects the usefulness of the buttons.

Given that a button can be in only two states (pressed or released), then the button's `is_pressed()` method returns a Boolean value. Sometimes you need to know if a button was pressed while the device was doing something else (such as scrolling text along the display). In this case, use the `was_pressed()` method of a button object to return a Boolean indicating if it had been pressed since the device started or the last time the method was called. Finally, you may need to count the number of presses for a button. Use `get_presses()` to return the running total and reset the total to zero.

These three methods give you a remarkable amount of flexibility when using the micro:bit's buttons. However, they hide some of the lower-level details of how such buttons work.

Buttons, in general, are digital in that they can only ever be either on or off. Obviously the microcontroller needs to detect the on or off state via the pin to which the button is connected. To differentiate the state, the microprocessor measures the voltage into the pin connected to the button.

A pin can be in three possible states: high, low, or floating.

If the pin detects some arbitrary signal, for example, 3.3 volts, then it is high; whereas if it can't detect a signal (0 volts), then it is low. In order for the pin to behave in a well-defined manner under all conditions, it is necessary to set the pin to be either pulled up (where the default signal is high) or pulled down (where the default signal is low). If we don't do this, the pin will be in the floating state: the microcontroller may unpredictably interpret the input as either high or low. By setting the pull we are setting, at the hardware level, a default value.

Armed with this information about the fundamental properties of pins, let's examine how the Circuit Playground Express works with its two buttons. The following simple script demonstrates working with buttons at a lower level than on the micro:bit (where such implementation details are hidden from the user). It creates a light display: press the lefthand button to make the NeoPixels flash clockwise or press the righthand button to reverse direction:

```
import neopixel
import time
import digitalio
from board import NEOPIXEL, BUTTON_A, BUTTON_B


np = neopixel.NeoPixel(NEOPIXEL, 10, auto_write=False)
button_a = digitalio.DigitalInOut(BUTTON_A)
button_a.pull = digitalio.Pull.DOWN
```

```
button_b = digitalio.DigitalInOut(BUTTON_B)
button_b.pull = digitalio.Pull.DOWN


clockwise = True
while True:
    time.sleep(0.05)
    if button_a.value:
        clockwise = True
    elif button_b.value:
        clockwise = False
    for i in range(10):
        if clockwise:
            i = 9 - i
        for j in range(10):
            np[j] = tuple((max(0, val - 64) for val in np[j]))
        np[i] = (0, 0, 254)
        np.write()
```

This script should look familiar—it's similar to the example used to demonstrate Neo-Pixels in the previous chapter. The most immediate difference is the use of the `digitalio` module to create two objects representing buttons A and B.

Both buttons are instantiated as `DigitalInOut` objects that represent digital pins that can act as both input or output. The button objects (`button_a` and `button_b`) are instantiated with a reference to the pin to use on the actual board. These are the `BUTTON_A` and `BUTTON_B` constants imported from the `board` module. The default state of such objects is to read input (this can be changed with the object's `switch_to_output` and `switch_to_input` methods), so the default works for the purposes of this example. Once instantiated, the button objects have their pull set to `DOWN`. This means the default signal will be low (i.e., the same as `False` in Python), which makes sense, since a button's default state is released: we only want the button to be "on" (or `True` in Python) if it is pressed. To get the state of the button, one simply reads the `value` attribute.

With this in mind, the example script sets the `clockwise` flag depending on which button has been pressed.

The Circuit Playground Express also has a switch that's similar to a button insofar as you move it with your fingers. However, a switch remains in the state to which you set it, rather than reverting to a default state once you release it. The switch on the Circuit Playground Express can be in one of two states just like the `clockwise` flag in the original script. As a result, we could re-write the example to use the switch as follows:

```
import neopixel
import time
import digitalio
```

```
from board import NEOPIXEL, SLIDE_SWITCH


np = neopixel.NeoPixel(NEOPIXEL, 10, auto_write=False)
switch = digitalio.DigitalInOut(SLIDE_SWITCH)
switch.pull = digitalio.Pull.UP


while True:
    time.sleep(0.05)
    for i in range(10):
        if switch.value:
            i = 9 - i
        for j in range(10):
            np[j] = tuple((max(0, val - 64) for val in np[j]))
        np[i] = (0, 0, 254)
        np.write()
```

I will leave it as an exercise for the reader to work out how it works, although everything you need to understand the script was explained when describing the button-based version.

The PyBoard also has a button labelled USR (in addition to the reset button) and takes a slightly different approach to the micro:bit and Circuit Playground Express. For a start, and rather confusingly, it calls the button a switch. As a result, it's controlled via a Switch object:

```
import pyb

led = pyb.LED(1)
sw = pyb.Switch()
while True:
    pyb.delay(100)
    if sw():
        led.toggle()
```

This fragment of code is very close to how we treated buttons on the micro:bit and Circuit Playground Express (although the implementation details are different because we are using a version of MicroPython for a different device). As before, there is an event loop and delay. The Boolean value of the switch is determined by calling the sw object that represents the switch. From the user's point of view, if you press the button, it toggles the PyBoard's red LED on and off.

However, MicroPython on the PyBoard provides an interesting alternative way to interact with buttons through the use of a *callback*. A callback is a function that's called when a certain event happens (such as a button press). To identify when a certain event has occurred, MicroPython uses an *interrupt*. An interrupt is simply a signal that something needs immediate action. In this case, MicroPython sets up an interrupt trigger on the pin to which the switch is connected.

When the button is pressed, the pin changes state from low to high, causing the microcontroller to register the change. It pauses what it's doing by saving its current state and calls the interrupt handler associated with the button. The interrupt handler executes the callback function, and the microcontroller is notified that the interruption has been handled. At this point, the microcontroller restores its pre-interrupt state and continues as before. The code that was running doesn't notice that it was interrupted.

Using this interrupt/callback method, we can simplify the LED toggling code:

```
import pyb


sw = pyb.Switch()

def my_callback():
    pyb.LED(1).toggle()

sw.callback(my_callback)
```

When the button is pressed, the `my_callback` function is called. This will interrupt any other code running at that moment in time. In case you were wondering, if more than one interrupt fires at the same time, then the one with the highest (pre-ordained) priority takes precedence, followed by any others in order of their priority. The interrupt for the button is set at the lowest priority.

To clear a callback, simply set it to None like this: `sw.callback(None)`.

There's one more type of finger-related interaction you can perform with the micro:bit and Circuit Playground Express: capacitative touch. Because the human body has quite a large capacitance (i.e., the ability to store electric charge), it's possible to detect a change in the capacitance of the pin and whatever is connected to it. If you are touching a pin, you are connected to it, and it's possible to detect differences due to the capacitance of your body.

This is only possible on the micro:bit with the large pins labelled, 0, 1, and 2. On the Circuit Playground Express, all the non-power or ground pins can detect capacitative touch.

As one might expect on the micro:bit, detecting touch is very simple:

```
from microbit import display, Image, pin0


while True:
    display.show(Image.ASLEEP)
    if pin0.is_touched():
        display.show(Image.HAPPY)
```

The is_touched method returns a Boolean to indicate if it's being touched. In the previous example, if pin 0 is touched, the sleeping face shown on the display is changed to a happy face. The Circuit Playground Express is only a little more complicated, but the effect is far more interesting:

```
import neopixel
import touchio
import digitalio
from board import *

# Stops the speaker crackling when touched.
spkr = digitalio.DigitalInOut(SPEAKER_ENABLE)
spkr.switch_to_output()
spkr.value = False


np = neopixel.NeoPixel(NEOPIXEL, 10, auto_write=False)
touch_a1 = touchio.TouchIn(A1)
touch_a3 = touchio.TouchIn(A3)
touch_a4 = touchio.TouchIn(A4)
touch_a6 = touchio.TouchIn(A6)


while True:
    if touch_a4.value:
        np[0] = (255, 0, 0)
        np[1] = (255, 0, 0)
    if touch_a6.value:
        np[3] = (0, 255, 0)
        np[4] = (0, 255, 0)
    if touch_a1.value:
        np[5] = (255, 255, 0)
        np[6] = (255, 255, 0)
    if touch_a3.value:
        np[8] = (0, 0, 255)
        np[9] = (0, 0, 255)
    for j in range(10):
        np[j] = tuple((max(0, val - 32) for val in np[j]))
    np.write()
```

The end result is differently coloured NeoPixels light up if an adjacent pin is touched. To make the effect feel more "alive", the NeoPixels gradually dim when you stop touching. As with the buttons on the Circuit Playground Express, you have to instantiate an object to represent the pin in the right sort of a way—in this case, we use the TouchIn class found in the touchio module. As is always the case, you need to pass in a reference to the physical pin via an object imported from the board module (in this case, the objects, A1, A3, and so on). Inside the event loop are some conditionals to check if the pin objects are registering high. If they are, the appropriately close NeoPixels are lit.

Pin A0 is also attached to the speaker. If touched (and this is likely in this example), it will cause the speaker to crackle. This behaviour is undesirable, so towards the start of the code a DigitalInOut object is created with reference to the SPEAKER_ENABLE pin. Setting the output of the resulting spkr object to False turns off the speaker to solve the crackling speaker problem.

This example reminds me of the classic electronic game *Simon Says*. How could you create a capacitive touch version with the Circuit Playground Express?

If the LCD screen is attached to the PyBoard, it can also be used to respond to touch. The most useful methods are is_touched, which returns a Boolean to indicate if the screen is currently touched, and get_touch, which returns a tuple of three values representing active (i.e., the screen is currently being touched), X and Y (the coordinates of the touch). With only a few lines of code, it is possible to create a simple finger-painting program:

```
import lcd160cr

lcd = lcd160cr.LCD160CR('X')
lcd.erase()
while True:
    a, x, y = lcd.get_touch()
    if a:
        lcd.set_pixel(x, y, lcd.rgb(255, 255, 255))
```

While it's hard to be accurate when painting on such a small device, and you're only limited to black and white (how would you improve the script?), I believe the results are quite impressive for only seven lines of code (see Figure 8-1).

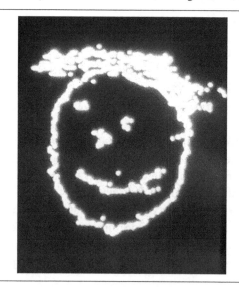

Figure 8-1. A presidential portrait, but of which president?

Accelerometers, Gestures, and Compasses

When you use a modern mobile phone, it is able to detect how it is oriented, the direction it is pointing, and sometimes how to react to gestures (such as a shake to cancel a certain operation). This is remarkably useful as a means of user interaction. For example, as you rotate your phone, the display is flipped from portrait to landscape mode. Certain gestures may also indicate certain states, such as the aforementioned shaking or placing the phone face down to turn off audible alerts. Finally, if the phone can detect its heading, this information can be used in conjunction with GPS signals in a mapping application to give you directions or, on its own, so your phone becomes a compass.

Such attributes rely on readings from relatively simple components: the accelerometer and magnetometer (compass).

An *accelerometer* is an electromechanical device consisting of a mass on a spring. As the mass moves in a certain direction due to force of gravity or an acceleration,[1] the capacitance changes between the moving mass and a fixed plate, allowing the mechanical movement of the mass to be represented by changes in electrical current. As with the touch-related interactions described, we use capacitance to measure things, be it the capacitive properties of the human body (for touching pins) or the movement of a mass adjacent to a fixed plate to measure gravity.

The accelerometers used in the PyBoard, micro:bit, and Circuit Playground Express actually consist of three seperate sensors to detect gravitational force along three perpendicular axes called X (left to right), Y (forwards and backwards), and Z (up and down).

In contrast, a magnetometer measures magnetic fields (a compass is a very simple example of this sort of device). In the case of the magnetometer on the micro:bit, it uses a miniature Hall-effect sensor that detects the Earth's magnetic field along the same X, Y, and Z axes as the accelerometer. The sensor produces voltage proportional to the strength and polarity of the magnetic field along each axis. This, in turn, is converted to digital signals representing the magnetic field intensity along each axis. By calibrating the compass and taking measurements along these axes, it is possible to determine a heading (turning the micro:bit into a compass). Alternatively, such measurements can be used as a very basic metal detector.

The APIs for the accelerometers on the PyBoard, micro:bit, and Circuit Playground Express are very similar: you get back measurements for the X, Y, and Z axes. For example, the following fragment of code for the micro:bit uses the accelerometer to

---

1 Einstein famously showed that these two effects are equivalent.

steer a pixel with a glowing tail around the display (moving pixels around on the screen like this is a fundamental feature of many games on the device):

```python
from microbit import *

x = 2
y = 2
sensitivity = 50
pause = 90
fade = 2

while True:
    roll = accelerometer.get_x()
    yaw = accelerometer.get_y()
    if roll < -sensitivity:
        x = max(0, x - 1)
    elif roll > sensitivity:
        x = min(4, x + 1)
    if yaw < -sensitivity:
        y = max(0, y - 1)
    elif yaw > sensitivity:
        y = min(4, y + 1)
    for i in range(5):
        for j in range(5):
            brightness = max(0, display.get_pixel(i, j) - fade)
            display.set_pixel(i, j, brightness)
    display.set_pixel(x, y, 9)
    sleep(pause)
```

The important lines are where we call `accelerometer.get_x` and `accelerometer.get_y`. There is also a notion of `sensitivity` such that movement won't register in a direction until some threshold is reached (in the example, this is given the arbitrary value 50, arrived at through experimentation). This is because the accelerometer is very sensitive, and humans do not have such fine-grained motor skills to work with such sensitive devices. The threshold means we have to tip the device in one direction or another more than enough for us humans to be able to register the difference. When some aspect of the hardware is used to interact with humans, there often needs to be some sort dampening of sensitivity via thresholds or bucketing values so the device is both manageable and usable.

In the case of an accelerometer, such interactions are reminiscent of the Wii remote and other similar peripherals for console games used to control player characters and other game-related assets. The tilting and movement of an accelerometer is a useful metaphor for controlling an aspect of some other thing (such as the position of a pixel). An accelerometer is also useful if you need to log the forces applied to an object, such as a model rocket.

Another use of the accelerometer is to detect gestures, such as shake, freefall, or face up. Such gestures are useful from a user interaction point of view since they can also represent states or instructions. For example, shake may mean something negative like "cancel and restart the game"; freefall is probably an indication that the device is in the process of being dropped, so prepare for a crash landing; and face up may just mean display something (since you look down upon the display of the micro:bit) in a similar manner to the way mobile phones activate their screens when you take them away from your ear to look at them.

Currently, only the micro:bit has built-in support for gestures. The following example demonstrates how they can be used:

```
from microbit import *

while True:
    if accelerometer.was_gesture('shake'):
        display.show(Image.ANGRY)
    elif accelerometer.was_gesture('face up'):
        display.show(Image.ASLEEP)
    elif accelerometer.was_gesture('up'):
        display.show(Image.HAPPY)
    sleep(100)
```

If the device is shaken, it displays an angry face; if it is flat but face up, it appears asleep; and if it is held upright, it's happy to see you. The micro:bit can recognise the following list of gestures: up, down, left, right, face up, face down, freefall, 3g, 6g, 8g, and shake.

The micro:bit is the only device with an onboard magnetometer. It's not very accurate and requires calibration before use. Calibration is achieved via the compass.cali brate method, which causes the device to wait until you've drawn a blocky circle on the display by rotating the device to move a pixel around the screen. Once calibrated, the micro:bit is able to report a heading with 0 as "north" (or some other strong magnetic field). Here's how to turn the device into a compass that displays where the micro:bit thinks north is:

```
from microbit import *

compass.calibrate()
while True:
    sleep(100)
    needle = ((15 - compass.heading()) // 30) % 12
    display.show(Image.ALL_CLOCKS[needle])
```

Sound, Light, and Temperature

Sound, light, and temperature-sensing roughly translate to the human senses of hearing, sight, and touch (although, such sensing is nowhere near as good as the human equivalent). The Circuit Playground Express has such sensors built in.

Sound sensing is done with a *microphone*, a device that turns sound waves into electrical signals. In the case of the part on the Circuit Playground Express, it's another small electromechanical device that works in a way that is similar to the accelerometer: there's a membrane etched into the device that vibrates in response to the changes in air pressure caused by sound waves. As the membrane vibrates, the capacitance between the membrane and a fixed plate changes, allowing the mechanical movement of the vibrating membrane to become changes in electrical current and eventually a digital signal.

Light sensing is essentially an LED in reverse. If you remember, when an electrical current is is applied to an LED, an electroluminescent semiconductor emits light. However, if light is shone onto an LED, an electrical current will flow through the LED but in the opposite direction to the flow when the LED emits light. By measuring this current we can use an LED to detect light. This is how light detection works on the micro:bit. The light-detecting capabilities of the Circuit Playground Express are provided by a phototransistor built into the board. A phototransistor is a component that specialises in detecting light (rather than re-using the characteristics of LEDs) and, just like the reverse LED trick, turns light energy into electrical current.

Temperature sensing is done with a *thermistor*, a component whose resistance changes with temperature. This is how the temperature sensor on the Circuit Playground Express works. However, many chips, including the microcontrollers that run MicroPython, have a minute on-chip thermal diode that's used to monitor the temperature of the chip. A thermal diode changes voltage across it according to temperature: as the temperature increases, the voltage decreases. This change is used to measure the temperature of the chip, and it is how temperature is read on the micro:bit.

Given the physical properties of such sensors, how can we use them?

The following example of how to use the microphone on the Circuit Playground Express makes use of an API that is not yet available at the time of publication.

However, given the expected speed of development, chances are the API will be in the latest version of CircuitPython very soon (look for a version 2.0 or greater).

While the following example demonstrates the basics, the final version will also include more capabilities, such as streaming audio data onto the filesystem, so you'll be able to record much longer fragments of sound.

The simplest way to use the onboard microphone of the Circuit Playground Express is to record short snippets of audio into a buffer. The following example is a magic echo machine. When the device starts, the NeoPixels around the edge of the board sequentially light up to indicate a sort of countdown. When the microphone is listening, all the NeoPixels are at a full green brightness. You have about a second's worth of time. The NeoPixels switch off, and the device plays back what it recorded via the onboard speaker. If you hold down button A during this process, the device will play back the recording at "chipmunk" speed. Holding down button B has the opposite effect: the audio is slowed down and lowered to mimic Barry White. Of course, if you don't press any of the buttons, you'll hear the audio played back at the correct pitch.

```python
import neopixel
import audiobusio
import digitalio
import audioio
import time
from board import *


def countdown(np):
    """ Uses the NeoPixels to display a countdown."""
    # Start from an "off" state.
    np.fill((0, 0, 0))
    np.write()
    for i in range(10):
        np[i] = (0, 20, 0)
        np.write()
        time.sleep(0.5)
    np.fill((0, 128, 0))
    np.write()


def record():
    """ Returns a buffer of recorded sound."""
    buf = bytearray(8000)
    with audiobusio.PDMIn(MICROPHONE_CLOCK, MICROPHONE_DATA) as mic:
        mic.record(buf, len(buf))
    return buf


def play(buf, freq):
    """
    Play the referenced buffer of recorded sound at a certain
    frequency.
    """
    # Set the speaker ready for output.
    speaker_enable = digitalio.DigitalInOut(SPEAKER_ENABLE)
    speaker_enable.switch_to_output(value = True)
    # Play the audio buffer through the speaker.
    with audioio.AudioOut(SPEAKER, buf) as speaker:
        speaker.frequency = freq
        speaker.play()
        # Block while the speaker is playing.
        while speaker.playing:
```

```
        pass

    neopixels = neopixel.NeoPixel(NEOPIXEL, 10, auto_write=False)
    button_a = digitalio.DigitalInOut(BUTTON_A)
    button_a.pull = digitalio.Pull.DOWN
    button_b = digitalio.DigitalInOut(BUTTON_B)
    button_b.pull = digitalio.Pull.DOWN


    countdown(neopixels)
    audio_buffer = record()
    neopixels.fill((0, 0, 0))
    neopixels.write()

    freq = 8000  # Default = normal speed.
    if button_a.value:
        freq = 12000  # Button A = chipmunk.
    elif button_b.value:
        freq = 6000  # Button B = Barry White.

    play(audio_buffer, freq)
```

The block of code of interest to us is in the record function. A bytearray buffer is created and used by an instance of the PDMIn class found within the audiobusio module. The class is instantiated with references to the microphone clock and data pins needed to perform any recording. The "PDM" in PDMIn is *pulse density modulation*, a method of representing an analog signal with binary (on/off) data. The relative density of pulses in the binary data corresponds to the analog signal's amplitude. Put (very) simply, a higher density of "on" values occurs at the peaks of a wave, whereas a lower density occurs in the troughs. In any case, the resulting mic class has a record method that fills the buffer buf with bytes representing recorded sound measured using pulse density modulation. At the end of the function, the buffer is returned for further processing.

The record function, used in concert with the countdown and play functions, turns the Circuit Playground Express into a silly sound-based toy. (We'll cover the speaker on the Circuit Playground Express in some detail in Chapter 11.)

From a programmatic point of view, the light sensor is used in a different way to the microphone: it is an analog input pin whose value relates to the amount of light detected by the physical sensor. The higher the number, the more light is detected. The same is true of the temperature sensor, with the resulting numbers reflecting changes in the temperature. The following REPL session demonstrates:

```
>>> import analogio
>>> from board import *
>>> light = analogio.AnalogIn(LIGHT)
>>> light.value  # in a dark place
```

```
152
>>> light.value  # held up at daylight
12037
>>> temp = analogio.AnalogIn(TEMPERATURE)
>>> temp.value  # ambient room temperature
30075
>>> temp.value  # after blowing warm air on the sensor
36405
>>> temp.value  # waiting a few seconds for it to cool down
34208
```

You are probably wondering how such raw analog readings from a pin can be turned into something useful. While such cute REPL-based demonstrations illustrate a point, they are not that useful in terms of getting a reading expressed in a meaningful unit of measurement. For this to happen, you will need to use the libraries currently in development (*https://github.com/adafruit/Adafruit_CircuitPython_Bundle*) created by Adafruit. Development is ongoing and fast moving (hence my reticence to write about them at this moment in time since they are likely to change); however, using them is as simple as downloading the latest release and copying the modules over to the flash-based filesystem of the Circuit Playground Express.

A quick example will suffice to demonstrate such libraries in action. Here's how to get the current temperature in degrees Celsius from the Circuit Playground's thermistor:

```
>>> import adafruit_thermistor
>>> import board
>>> thermistor = adafruit_thermistor.Thermistor(board.TEMPERATURE, 10000, 10000,
                                                 25, 3950)
>>> thermistor.temperature
26.60413
```

The arguments used by the `Thermistor` class relate to settings dependent on the model of thermistor in use. The ones used in the preceding example are correct for the Circuit Playground Express.

In the case of the micro:bit, the temperature is expressed in degrees Celcius. It represents the current temperature of the microcontroller rather than the ambient temperature:

```
>>> from microbit import temperature
>>> temperature()
24
```

Sensing with Peripherals

While the micro:bit, Circuit Playground Express and, to a lesser extent, PyBoard have inputs and sensors built into the boards, it is possible to connect such peripherals to the GPIO pins of any of the devices running MicroPython. The important thing to remember is that you access the device in exactly the way that has been demonstrated throughout this chapter: via pins. In the case of external peripherals, they will use the externally available pins rather than "pins" directly attached to built-in components attached to the board.

A very simple example using ESP8266-based boards (as yet, unused in this chapter) will be sufficient to demonstrate the general principal (see Figure 8-2).

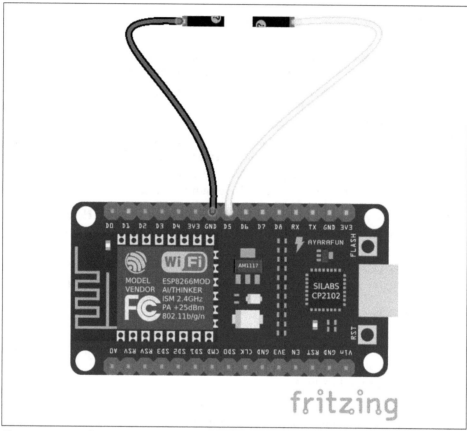

Figure 8-2. Simple sensing with two wires

Very carefully connect wires to the GND and pin labelled D5 on the physical board. We're going to simulate a simple digital signal that could be created by an external button. The following code illuminates the onboard LED every time you touch the wires together:

```
from machine import Pin

led = Pin(2, Pin.OUT)
button = Pin(14, Pin.IN, Pin.PULL_UP)
while True:
    led.value(button.value())
```

Both the `led` and `button` objects are instances of the `Pin` class, which is instantiated with a pin number and an indication of whether it is to be used for output (as the `led` is) or input (as the `button` is). Since we need to take a reading from the `button`, we ensure that it's not in a floating state by passing in a third `PULL_UP` argument. Finally, the value set for the LED is whatever the value of the button is read to be. As the wires touch, changing the value of the input into the `button` pin, so the output is changed to the `led` pin.

Given such playful experiments with both visual output, inputs, and sensors, we are in a good position to take a detailed look at not only how the GPIO pins work but also at the various protocols you might use to communicate with attached devices.

## CHAPTER 9

GPIO

General Purpose Input and Output (GPIO) is how all the devices connect to the external world.

This connection is achieved in a physical sense via "pins" that ultimately connect to the microcontroller running MicroPython. By controlling or reading the voltage from the pins, MicroPython is able to both sense and control the external world through the peripherals connected to them. Each pin is given a name so we can reference it and, depending on how it is configured, is capable of processing and emitting different sorts of signals.

This chapter explains how pins work and describes three common protocols that use the pins to communicate with the outside world: UART, SPI, and I²C. Such protocols make interacting with external peripherals both easy and standardised.

Pins

"Pins" is a generic term for things that, historically, looked like pins but these days, often do not. For the purposes of this book, a pin is a conductive area connected to the microcontroller through which communication may take place with external peripherals. Figure 9-1 shows a close-up picture of the "pins" on the micro:bit:

Figure 9-1. Pins on a micro:bit

They don't look like pins at all, and some of them are big enough for you to attach an alligator clip. The pins form the bottom edge of the board, and you may be wondering how you are supposed to connect things to all the smaller pins. The answer is to use an edge connector into which you plug jumper cables connected to external peripherals or a breadboard onto which you can place external components (Figure 9-2).

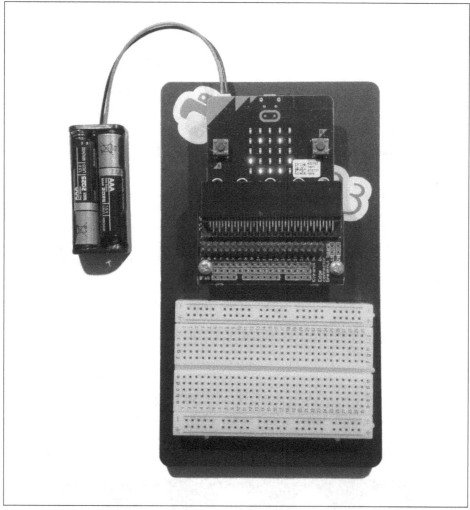

Figure 9-2. A micro:bit in an edge connector attached with an adjacent breadboard

The pins on a Circuit Playground Express are all like the micro:bit's—large for alligator-clip-related reasons. In contrast, the PyBoard comes in two configurations: without any pin connections (there are just holes in the circuit board into which one solders such connectors) or pre-soldered with female pins into which one pokes jump cables to which you attach the external peripherals. The ESP8266/32-based boards often come with male pins pre-soldered onto the board—at last, devices with GPIO pins that actually look like pins!

Pins are named so we can reference them in our code. References to pins are found in different places, depending on the version of MicroPython you have running on your device. If you're using the micro:bit you'll find them in the `microbit` module. They're in the `board` module if you're using CircuitPython with Adafruit devices. Both the original PyBoard and the ESP8266/32 ports of MicroPython have a `Pin` class that you instantiate with the name of the pin and some notion of its characteristics (for example, that it's a digital input).

Names are usually printed onto the board so it is possible to look at the pin and work out what it's called. Different pins may be used for difference sorts of things. Some pins simply provide electrical current at an advertised voltage in order to power an external peripheral. The "3v" pin on the micro:bit is a good example of this sort of pin, and you can think of it as the equivalent of the positive end of a battery.

Other pins act as *ground* (often labelled GND), which is the equivalent of the negative end of a battery. The pins that only provide current and those labelled ground are not under your control since they only do what their name suggests.

It is the other pins that are more interesting to us and they may be capable of doing different sorts of things. For instance, all of them will be capable of acting as digital pins. You control them to be either low (0 V) or high (producing current at the board's supply voltage, often 3.3 V). Some others will be able to act as analog pins, capable of sending or receiving signals that are not high or low but may be somewhere between each extreme. Usually such graduation in value is manifested as differences in voltage that are read by an analog-to-digital converter (ADC) and turned into a number within a certain range. Analog output is created by a digital-to-analog converter (DAC) that takes a number and turns it into a voltage representation of the analog value. Of course, digital pins can pretend to be analog by using the pulse width modulation trick described Chapter 7. Finally, some pins are configured in such a way as to allow them to respond to capacative touch (as described in Chapter 8).

Remember that GPIO pins can be in three default input states: high, low, and floating. By setting the "pull" of the pin (to high or low), we avoid the indeterminate floating state whose signal will reflect the ambient electrical conditions of the pin.

On some boards, it is possible to define interrupts that kick in if their input changes. This generally follows the pattern of defining a simple callback function to handle the interrupt and assigning it to a type of change on a specific pin. The following example for the ESP8266 boards demonstrates this:

```
from machine import Pin

def callback(p):
    print('Pin', p)

p0 = Pin(0, Pin.IN)
p0.irq(trigger=Pin.IRQ_FALLING, handler=callback)
```

The `callback` function receives an object representing the pin, p, that triggered the interrupt and prints it. Such hard interrupts trigger as soon as the expected event occurs, interrupting any running code. As a result, the callback functions that handle such interrupts are limited in what they can do (for example, they cannot allocate memory) and should be as simple as possible.

Next, an input pin is defined, and we assign the callback function as a handler for an interrupt request (IRQ) by defining the trigger (the pin drops from a high to low state) and referencing the `callback` function. From this moment on, if you apply high and then low voltages to pin 0, you'll see the results of the `print` function used in the callback.

Sometimes you only need to use a single pin to send or receive a signal. This is called *serial communication*, since the data is sent sequentially, a single bit at a time. Alternatively, you may need to send or receive data via multiple pins. This is called *parallel communication*, as several bits are sent at once over the available channels of communication. Such connections that carry signals between devices and components are called a *bus*.

To work out what pins work in what way (if the name of the pin doesn't tell you that already), you should consult the pinout diagram for the device. Figure 9-3 shows what the micro:bit's pinout looks like.

Notice how each pin has a name and some indication of its function. Some of the pins are re-used to control things like the LEDs on the display. Rather than reproduce pinouts in this book (that with new iterations of boards may result in changes), I suggest you look online for them, typing the name of the device and the word "pinout" into a search engine.

While this sort of information is useful, many of the peripherals you will want to use with your boards use protocols that sit on top of the physical capabilities of the various GPIO pins. It is to three of these protocols that we turn our attention for the rest of the chapter. Once you understand the basics of each of these protocols, it should be a relatively simple task to connect a peripheral, read its associated data sheet (produced by the manufacturer), and work out how to use the expected protocol to make use of it.

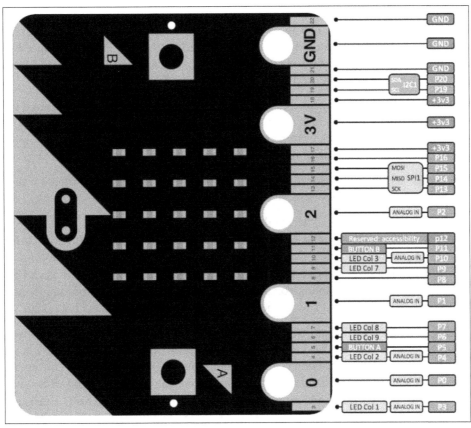

Figure 9-3. The micro:bit's pinout diagram

UART

When you plug a board into your computer via the USB cable it is possible to communicate with the device using the REPL. What makes that possible is the *universal asynchronous receiver/transmitter* (UART), a part of the microcontroller that mediates between serial and parallel communication. Serial messages come in one bit at a time (a high/low signal), and the UART hardware assembles the signal into *bytes* (a parallel representation usually consisting of 8 bits) that are sent via an internal bus for further processing by the microprocessor. Conversely, to send a message the UART takes a byte and turns it into a series of high/low signals representing the constituent bits.

For this to work, several arrangements need to be made. First, the transmitting port (usually called TX) of device A must connect to the receiving port (usually called RX) of device B, and vice versa.

Second, there also needs to be agreement about the timing of the serial communication so the UART can detect the individual high/low signals. This is the speed of communication and is expressed as one of several standard baud rates: 9600, 14400, 19200, 28800, 38400, 57600, and 115200 bits per second.

Third, sometimes you may need to specify the number of bits per byte (although the standard is usually 8). You may also need to specify whether to use a parity bit (whose function is to detect errors in transmission), and the number of stop bits that signal the end of a unit of transmission.

The UART also has a ""first in/first out" (FIFO) queue so bytes can be buffered if they are not read as soon as they're received.

By default, the UART on MicroPython boards is connected to the internal USB-UART TX/RX pins that connect to a USB serial convertor, thus connecting the UART to your PC via the USB port. On the PC end of things, a library like pySerial (*https://pythonhosted.org/pyserial/*) or a tool like picocom opens a serial connection via a USB port on your PC, thus enabling you to send and receive data to and from the Python REPL. The default baud rate for connecting to MicroPython in this way is 115200.

UART interactions in MicroPython require that the connection is configured (specifying the pins, baud rate, and other attributes already discussed). Each board has a slightly different way to instantiate and configure the UART although, at a conceptual level, they all work in the same way. Once configured, you will be working with a byte stream with familiar methods such as read, readline, and write. This is consistent across all platforms. The following micro:bit-based example is typical and demonstrates how to use the UART to read and write to a connected PC via the USB-serial bus:

```
from microbit import *

while True:
    msg = uart.read()
    if msg:
        uart.write(msg)
```

This short script simply echos anything it receives (it uses the default UART settings, so it is receiving and transmitting via the USB port). If you connect to the device in the same way you would with the REPL, it will just reply with any of the characters you type. It's a very basic example, but all the fundamentals are contained within the script: read from the buffer and write a response. It is important to note that the micro:bit has a uart object that mediates such communication. Other boards will require you to instantiate a UART class with the right configuration for your needs. In this case, please consult the documentation for the port of MicroPython that targets your device. It's also important to point out that UART isn't just for REPL- or USB-based interactions; it can be used to facilitate all sorts of useful yet simple inter-device communication.

SPI

As the name suggests, the *serial peripheral interface* (SPI) is another serial protocol whose aim is to facilitate communication with peripherals. However, it is different from using the UART in a number of important ways.

As you know, the UART is an *asynchronous protocol*, meaning there is no signal used to indicate timing synchronisation to an agreed single clock when communicating between devices. All each device knows is the expected baud rate (speed) of transmission that has been agreed in advance. However, this can be a problem if the two devices have slightly different clocks: if the receiver samples the signal at the wrong time (to ascertain the high or low state on the pin), it will end up producing garbage. To work around this problem, the transmitting UART will add bits (for example, the stop bit) to help the receiving UART synchronise with the data as it arrives. Differences in data rate are not usually a problem in this case because the receiver will re-synchronise upon receipt of the stop bit. However, such asynchronous communication adds a lot of overhead in the form of stop bits, and the relatively complicated UART hardware needed to make such communication possible. Sometimes we need to connect with relatively simple peripherals that may not have such capabilities built in.

SPI takes a different approach: it's a synchonous data bus, and there is a notion of hierarchy of devices.

SPI is synchronous because one of the connections between devices is an oscillating clock signal that tells devices exactly when to sample the high or low states of the signal (usually labelled as SCLK). As a result, the measures and complexity introduced to mitigate differences in clocks in UART-based communication are replaced by the clock signal.

You may wonder how devices tell where the clock signal comes from. This is answered by the hierarchical nature of SPI.

There is a primary device (usually the microcontroller) that, by prearrangement, supplies the clock signal. By convention, this is called the *master* with any other device connected via SPI referred to as *slave*[s].[1] In a similar way to how UART has TX and RX connections, the SPI protocol calls its data transmission connections *MOSI* (master out, slave in) and *MISO* (master in, slave out). All the slave devices receive and transmit on the same MOSI and MISO connections, so there needs to be some way to differentiate between signals to and from specific slave devices. This is achieved by the *chip select* (CS) connection (also sometimes called *slave select*). This connection indicates when a slave device should send and/or receive data and is done in an active-low configuration: the pins are pulled high by default and go low when they signal that the slave should activate. There are a couple of ways in which such slave-select signalling can be organised.

Figure 9-4 shows how the master device has a one-to-one CS connection with each of the slave devices. Each slave device is activated by its unique CS connection, although it means that the master must use as many separate pins as there are slave devices.

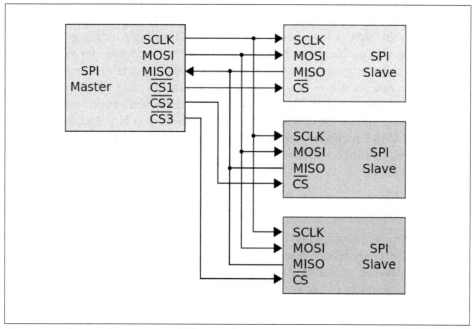

Figure 9-4. SPI configuration with three independent slaves

Some devices prefer to be in a "daisy chain" configuration. In this case, there is only one CS connection that simultaneously activates all the slave devices. However, the

---

1 I find the use of terminology such as "master" and "slave" distasteful (far better to say "primary", "secondary", or "tertiary", etc.), but it's the historic convention that's used in all the documentation that you'll read, so I'll hold my nose and continue to use such a convention in the hope that future engineers will name aspects of their protocols with a sympathetic appreciation of such loaded terms.

first slave's MISO is connected to the second slave's MOSI and so on so that all the slave devices are connected like a daisy chain. Data is transmitted by each slave by passing on, in the current group of clock pulses, an exact copy of the data received during the previous group of clock pulses. In this way the serial data eventually flows through to all the connected devices. When the CS connection activates all the slave devices, each device, by knowing its address in the sequence, processes the data at the appropriate location in the data series that flowed through the daisy chain. For example, if we had three devices, each expecting a byte of data, we send a sequence of three bytes that eventually flows through to all the devices (see). When the CS is activated, the first device uses the byte in position 0, the second device uses the byte in position 1, and the third device uses the byte in position 2. In this way, we can connect many devices together without using a large number of pins. This is called a *shift register* and is one way to convert serial communication into a parallel equivalent.

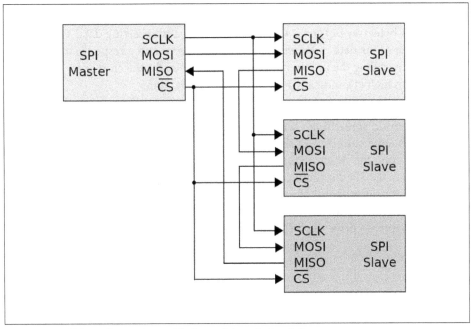

Figure 9-5. SPI configuration with three daisy-chained slaves

You will notice from pinout diagrams that some pins will be labelled with the names of SPI connections (SCLK, MOSI, MISO, etc.). Use these pins to attach SPI devices. Since there is an indeterminate number of CS connections, depending on how you are organising signalling, the CS pin should be selected and controlled by you. As with UART, each device is slightly different in terms of the steps to configure your SPI connections, although they are conceptually very similar. As before, you should consult the API documents for your device for the specific details. Nevertheless, the following REPL-based example for the PyBoard illustrates the essential steps you'll need to take to make things work:

```
>>> from machine import SPI, Pin
>>> spi = SPI('X')
>>> cs = Pin('X1', Pin.OUT)
>>> cs.value(0)
>>> buffer = bytearray(5)
>>> spi.write_readinto(b'hello', buffer)
>>> cs.value(1)
```

An `spi` object is instantiated with an indication of its position (there are two orientations for using SPI on the PyBoard: "X" and "Y"). The SPI class assumes the device will act as the master. A `cs` object is also instantiated from the `Pin` class to represent the chip select connection that is arbitrarily using the pin "X1" (see the PyBoard pinout to locate this pin). Note how it is set as an output. Pulling the `cs` object to low (0) indicates that any further interactions target the slave device connected to the chip select pin. A bytearray `buffer` is created for sending and receiving data. This must be the same size as the data to be sent. The message is sent and a response received into the `buffer` by calling the `write_readinto` method. Finally, the chip select connection is returned to high (1), indicating the end of the interaction.

In case you are wondering what bytes to send to SPI-connected peripherals in order to make them do something useful, or what their responses may mean, you should look at the manufacturer's data sheet for the device. It's important to note that, more often than not, the MicroPython community will have created a module that abstracts away the SPI communication for a particular device to allow you to concentrate on the "business logic" of your application rather than SPI-related implementation details. A good example is the `lcd160cr` module for the PyBoard that we used in an earlier chapter to drive the LCD colour display.

I²C

The inter-integrated circuit (I²C) protocol is the final hardware protocol you are likely to encounter with MicroPython.

Why another protocol? The answer can be found by looking at some of the drawbacks of UART or SPI.

With UART you are limited to a one-to-one connection and the overhead needed to mitigate problems with the asynchronous nature of the protocol. With SPI you require at least four connections (and potentially a lot more if you use several slave devices).

With I²C you only need two physical connections like UART, but those two wires could support many slave devices like SPI. Furthermore, I²C is capable of supporting a multiple master system whereby the master devices take it in turn to communicate with the slaves connected to the bus.

However, I²C isn't capable of SPI's speed of data transmission, so devices that need speedy data transfers will often use SPI rather than I²C. Both I²C and SPI are capable of much faster rates of data transfer than UART.

The two connections used by I²C are called *SCL* (for the clock signal) and *SDA* (for the data). The clock signal always comes from the current master. The connections for I²C are *open drain*, meaning they are pulled high by default, and devices change the signal by pulling low. This avoids the potential collision of a device driving the signal high while another is trying to pull it low (eliminating a situation where conflicting devices may cause damage to each other).

The protocol sent down the wire is more complicated than UART or SPI. Messages contain two parts: an address frame that indicates for whom the subsequent data is intended, and one or more data frames containing the actual data.

Communication is initiated by the master leaving the SCL connection high while pulling SDA low. This alerts the slave devices that a new message is about to be sent. The address frame is always the first thing sent in a new message and consists of a 7-bit address identifying the target slave, followed by a bit to indicate if the message is a read (1) or write (0). A ninth bit is used to allow the target slave to acknowledge receipt of the address frame. To do this, the target slave must pull the SDA line low before the ninth clock pulse. If this acknowledgment doesn't happen, the exchange halts, and it's up to the master to decide how to proceed.

Assuming a successful acknowledgment of the address frame, the subsequent data frames are synchronised by the master sending clock pulses via SCL, with the actual data transmitted via SDA by either the master or target slave (depending on if this is a read or write operation). The number of subsequent data frames is arbitrary in length and only stops when the master generates a stop condition: first SCL moves from low to high and remains high while SDA also moves from low to high.

While this may sound more complicated than SPI in terms of implementation details (it is), the end result is very easy to use from a programmatic point of view. Again, the caveats about different boards working slightly differently whilst being conceptually similar apply. Here is a REPL session as if using a PyBoard. The example assumes the connections as per the board's pinout diagram for SCL and SDA connections:

```
>>> from machine import I2C
>>> i2c = I2C('X')
>>> i2c.scan()
[46]
>>> i2c.writeto(46, b'A')
1
>>> i2c.readfrom(46, 8)
```

```
b'\x00A\x00A\x00A\x00A'
>>> i2c.writeto_mem(46, 0, b'A')
>>> i2c.readfrom_mem(46, 0, 2)
b'\x00A'
```

In a fashion similar to the SPI example earlier, an i2c object is instantiated with an indication of the orientation. The I2C class assumes the device will act as a master. Since there could be several devices connected to the I²C bus, it's possible to scan for their addresses (yielding a single device with the address 46). All subsequent communication uses this address to indicate the target device for instructions. The writeto and readfrom methods work as expected (sending a byte representation of "A" and receiving 8 bytes in return). The remaining two lines demonstrate how to write to and read from a particular memory address (indicated by 0) on a specific device (with the address 46).

As with SPI, you probably won't need to directly use I²C since MicroPython will provide modules for specific devices that wrap all the implementation details. However, if you do need to drop to the I²C level, you should consult the data sheet from the manufacturer of the attached peripheral to discover what messages are used to interact with the device.

Miscellaneous GPIO Techniques and Protocols

The topics covered so far in this chapter give you a good foundation of knowledge about GPIO and should cover most of your GPIO-related interactions. However, sometimes you will encounter something a little more esoteric, so this final section will examine some of the darker corners to cover such cases.

Bit banging sounds like a nerdy version of whack-a-mole. It's actually a lot more fun: it's when you ignore hardware protocols and use software to control pins, timing, levels, and synchronisation in order to create a low-cost, highly bespoke solution to a peripheral-related problem. To say that it's a "hack" is an understatement, but that's what makes it such fun. So if you hear of someone mentioning bit banging, what they really mean is that they're going "off piste" with code in terms of interactions with the hardware. Given the usually bespoke or experimental nature of a bit banging hack, the only common ground to describe here is that everything is controlled by the software, as needs apply, be that sampling, timing, controlling the signal, buffering, and so on. It's the embedded version of poking with a stick to see what happens until you've figured it out. Why would you do such a thing? Sometimes it's the simplest solution and reduces the overhead of complex code or allows you to abandon a large library (where you're using only a small subset of its capabilities) in order to save precious space or resources.

Some peripherals don't use UART, SPI, or I²C and have their own bespoke protocol. NeoPixels (also known as ws2812) and the digital humidity and temperature (DHT) line of sensors all use a 1-wire interface. These are just a couple of common examples you may run across and the actual implementation details are unimportant since MicroPython already has modules for both types of device. The point is that sometimes you may encounter a peripheral where none of the standard protocols apply and, if there's no module already available, you may need to roll up your sleeves and get bit banging to make the thing blink or go bloop.

Perhaps the most enjoyable aspect of programming GPIO pins with MicroPython is that you're close to the hardware. As Python programmers we're used to working in relatively abstract computing environments where Python collaborates with the operating system to make things work. In contrast, as embedded developers, working in MicroPython retains the closeness to the hardware while giving us a high-level, expressive, and easy-to-use language that allows us to create working solutions in only a fraction of the time it would take in other, lower-level languages. The lack of abstractions and simplicity combined with the expressiveness of Python is a clue as to why MicroPython is such an exceptional teaching tool: you're close enough to the hardware to be mucking about with how a computer actually works (rather than sitting on top of layers upon layers of abstractions), yet you have a powerful, flexible, and (most importantly) easy-to-learn programming language. That the skills learned with MicroPython are easily transferrable to "regular" Python is a testament to the continuity of experience that the Python ecosystem provides.

## CHAPTER 10

Networking

Due to the small size of the devices upon which MicroPython runs, there is no room to add an ethernet socket. As a result, those devices that include capabilities for inter-device communication do so wirelessly.[1]

There is something strangely satisfying about interacting at a distance: it feels like magic. However, there's something much better than magic going on: physics! How such communication works is fascinating, and two types of communication will be covered in this chapter: infrared (on the Circuit Playground Express) and radio (on the micro:bit- and ESP8266/32-based devices).

Infrared (IR) communication is most commonly associated with television remote controls: slabs of buttons to be poked and pressed while pointing the device at the TV. Infrared works well over a short distance so long as there's a line of sight between the transmitter and receiver (hence the requirement that you point the remote at your TV). We can't see infrared light because its wave length is just below that of visible light.[2] There are many sources of infrared light in our environment: the sun, light bulbs, candles, and even our own bodies (our bodies radiate most of their heat in the infrared spectrum). To overcome the potential for interference from such ambient sources of infrared light, when sending a signal it is common to modulate it. In this instance, modulation means the signal is sent at a pre-agreed frequency, the most common being 38 kHz (although other frequencies are used). Put simply, the infrared transmitter blinks at 38,000 times per second. Such a signal is unlike other ambient

---

1 For the purposes of this book, Adafruit's Circuit Playground Express, the BBC micro:bit, and ESP8266/32-based boards

2 Many cameras on mobile phones can detect infrared light. If you point a TV remote at such cameras and press a button, you'll see it flicker if viewed via the screen of your mobile phone.

sources of infrared light and thus stands out to the receiver. The duration of the modulated infrared signals is used to encode data.

Adafruit's Circuit Playground Express has both an infrared transmitter and receiver built into the board. Not only does this mean the device can communicate with your TV or receive signals from your remote, but, as we shall see, it's possible to send signals between devices.

In contrast, radio-based interaction uses radio waves (a type of electromagnetic radiation, similar to visible light). These don't require a line of sight and work around corners, through walls, and over much further distances than infrared signals. A property of radio waves (such as the amplitude, phase, or frequency) is modulated by a transmitter in such a say that information is encoded and, thus, broadcast. When radio waves encounter an electrical conductor (i.e., an aerial), they cause an alternating current from which the information in the radio waves can be extracted and transformed back into its original form.

The micro:bit and ESP8266/32 devices use radio communication in different ways. Due to its educational roots, the micro:bit's networking is so simple that an 11-year-old would be able to construct a mental model for how it works. In contrast, the ESP8266/32-based boards have an onboard TCP/IP stack and communicate via IEEE 802.11 standards (i.e., normal WiFi), making them useful for building Internet of Things projects.

The hardware on both the BBC micro:bit and ESP32 is also capable of Bluetooth communication. Both boards have hardware for Bluetooth Low Energy (BLE), with the ESP32's hardware also capable of standard Bluetooth. Unfortunately, we won't cover Bluetooth (but it's important to know it could be a feature in the future). The BLE implementation for the micro:bit takes up 12 Kb of the device's 16 Kb of RAM and more than 100 Kb of the 256 Kb of flash memory, leaving no room for MicroPython to run. There are some brave souls working on an ultra memory-efficient BLE stack for MicroPython on the micro:bit, although it remains to be seen if their work will bear fruit.

Nevertheless, networking with all of the devices is fun simply because it feels like magic. The remainder of this chapter is an overview with working examples to get you started in the world of interconnetced devices, networking, IoT protocols, and MicroPython.

Circuit Playground Express Infrared

As mentioned, an infrared signal encodes data. It works by representing digital on and off values by the presence (or absence) of infrared light modulated at 38 kHz. But for the signal to mean anything, we must use a protocol for communication. There are a plethora of protocols created by manufacturers of infrared devices. They generally share a common modus operandi: the duration between changes in the signal is used to represent data. Adafruit's Tony DiCola has produced a fantastic video tutorial demonstrating how this works (*https://www.youtube.com/watch?v=TIbp7DzfOBM*) and shows how to capture TV remote signals and replay them, allowing your own embedded devices to control your TV.

This is similar to Morse code. *Morse code* defines how to send character-based messages via on/off signals of long or short durations. Long durations are called "dah" and are usually written as dashes (-), whereas short durations are called "dit" and written as dots (.). By combining dashes and dots, Morse defines a way to send characters. For example, the letter "A" is defined as .- (dit dah). Characters are usually separated by short pauses, sometimes written as a backslash (/).

In the following examples, I'll use the Morse code to send a simple text-based message between two Circuit Playground Express boards. It has no practical use except as a means of demonstrating how to send and receive infrared messages using Circuit-Python.

Obviously, the timing of the signal is important: we need to be able to tell a dot from a dash. Furthermore, we need to be able to detect different characters, words, and the beginning and end of the message. Therein lies the point of a protocol: to agree beforehand on the details of how such things are represented by the signal. In this case, some arbitrary timings can be used to represent all we need. If we take microseconds as a unit of measurement, then a "dit" can be represented by a duration of 1,000 microseconds and a "dah" by 2,000 microseconds. The boundary between characters can be represented by a duration of 4,000 microseconds and a word boundary by two durations of 8,000 microseconds each.[3]

Given this simple protocol, here's how to send "Hello World":

```
import array
import pulseio
import board
```

---

[3] Two durations of 8,000 microseconds are used because we detect durations when the signal changes from on to off and vice versa. If the signal is currently in an "on" state, then a single off duration of 8,000 microseconds is indistinguishable from no signal at all. By sending two durations, we can be certain the 8,000 microsecond duration is explicit.

```python
# A lookup table of morse codes and characters.
MORSE_CODE_LOOKUP = {
    "A": ".-",
    "B": "-...",
    "C": "-.-.",
    "D": "-..",
    "E": ".",
    "F": "..-.",
    "G": "--.",
    "H": "....",
    "I": "..",
    "J": ".---",
    "K": "-.-",
    "L": ".-..",
    "M": "--",
    "N": "-.",
    "O": "---",
    "P": ".--.",
    "Q": "--.-",
    "R": ".-.",
    "S": "...",
    "T": "-",
    "U": "..-",
    "V": "...-",
    "W": ".--",
    "X": "-..-",
    "Y": "-.--",
    "Z": "--..",
    "1": ".----",
    "2": "..---",
    "3": "...--",
    "4": "....-",
    "5": ".....",
    "6": "-....",
    "7": "--...",
    "8": "---..",
    "9": "----.",
    "0": "-----",
}


def encode_message(msg):
    words = msg.split(' ')
    message_buffer = []
    for word in words:
        message_buffer.extend([8000, 8000, ])  # Indicates a new word.
        for character in word:
            message_buffer.extend([4000])  # Indicates a new letter.
            for val in MORSE_CODE_LOOKUP[character]:
                if val == '-':
                    message_buffer.extend([2000])  # Indicates a dah.
```

```
        else:
            message_buffer.extend([1000])  # Indicates a dit.
    if words:
        message_buffer.extend([8000, 8000, ])  # Indicates end of message
    return array.array('H', message_buffer)



ir_led = pulseio.PWMOut(board.REMOTEOUT, frequency=38000, duty_cycle=2**15)
ir_out = pulseio.PulseOut(ir_led)
message = encode_message("HELLO WORLD")
ir_out.send(message)
```

The code is simple. Import the modules needed, create a lookup table for converting characters into Morse code, define a function to convert a string containing the message into a memory-efficient array of unsigned short integers used by the PulseOut class to send the signal, and then configure things to send the message.

The interesting functionality resides in the pulseio module. Since we're modulating the infrared signal, we use the PWMOut class to instantiate an object representing a PWM signal, at the expected frequency (38 kHz) and duty cycle, that is emitted from the pin representing the infrared transmitter (REMOTEOUT). The resulting object is used to create a PulseOut class whose send method actually causes the signal to be transmitted.

The encode function is rather basic. It splits the message into its constituent words and then splits each word into its constituent characters and appends the resulting durations (as specified in the protocol described earlier) to a list that's converted into the array used by the PulseOut class when sending the message.

Receiving the message is relatively similar in that there's a lookup table to convert Morse code to letters, and instead of using the pulsio module's PulseOut class, the receiving PulseIn class is used. However, there are some interesting complications that need to be factored into this code:

```
import array
import pulseio
import board
import time

# A lookup table of morse codes and characters.
MORSE_CODE_LOOKUP = {
    ".-": "A",
    "-...": "B",
    "-.-.": "C",
    "-..": "D",
    ".": "E",
    "..-.": "F",
    "--.": "G",
    "....": "H",
```

```python
    "..": "I",
    ".---": "J",
    "-.-": "K",
    ".-..": "L",
    "--": "M",
    "-.": "N",
    "---": "O",
    ".--.": "P",
    "--.-": "Q",
    ".-.": "R",
    "...": "S",
    "-": "T",
    "..-": "U",
    "...-": "V",
    ".--": "W",
    "-..-": "X",
    "-.--": "Y",
    "--..": "Z",
    ".----": "1",
    "..---": "2",
    "...--": "3",
    "....-": "4",
    ".....": "5",
    "-....": "6",
    "--...": "7",
    "---..": "8",
    "----.": "9",
    "-----": "0",
}


VALID_VALUES = (1000, 2000, 4000, 8000)


def normalise(raw):
    """
    A generator function that yields normalised items from the raw input.
    """
    for val in raw:
        rounded_val = round(val/1000) * 1000
        if rounded_val in VALID_VALUES:
            yield rounded_val


def get_character(tokens):
    """
    Given a list of tokens (Morse code dahs and dits represented as "-" and
    "."), return the related character or "?" if there's no match.
    """
    return MORSE_CODE_LOOKUP.get(''.join(tokens), "?")
```

```
def decode_message(normalised):
    """
    Given a source of normalised incoming values, returns a string
    representation of the message contained therein.
    """
    # Split the incoming normalised values into words, characters and tokens.
    words = []
    characters = []
    tokens = []
    for val in normalised:
        if val == 8000:
            # A new word.
            # Store away the old tokens and characters and reset state.
            if tokens:
                characters.append(get_character(tokens))
            if characters:
                words.append(''.join(characters))
            tokens = []
            characters = []
        elif val == 4000:
            # A new character.
            # Store away and reset the tokens of the previous character.
            if tokens:
                characters.append(get_character(tokens))
            tokens = []
        elif val == 2000:
            # A dah (represented as '-')
            tokens.append('-')
        elif val == 1000:
            # A dit token (represented as '.')
            tokens.append('.')
    return ' '.join(words).strip()


ir_in = pulseio.PulseIn(board.REMOTEIN, maxlen=512, idle_state=False)


while True:
    while len(ir_in) == 0:
        time.sleep(1)
    ir_in.pause()
    raw = [ir_in[i] for i in range(len(ir_in))]
    normalised = normalise(raw)
    msg = decode_message(normalised)
    if msg:
        print(msg)
    ir_in.clear()
    ir_in.resume()
```

The normalise function acts on the raw duration values recieved from the infrared
signal. Because of inaccuracies between the clocks on different devices, the incoming
durations won't be exactly 1,000, 2,000, 4,000, or 8,000, but they'll be very close. As a

result, `normalise` rounds the raw values to the nearest 1,000 microseconds. Given the normalised input, the `decode_message` function turns the durations into a string of characters, thus reconstituting the text-based message contained therein.

The `ir_in` object is an instance of the `PulseIn` class instantiated with a reference to the infrared receiver's pin, an indication of the maximum number of durations to store at one (in this case, 512) and a default idle state (upon starting to receive a new message, the first recorded signal will be the opposite state from idle).

Since there's no way to know when a message will arrive, an event loop is created that checks if there are any incoming values (and if there are not, it sleeps for a second). Upon receipt of a message, the infrared receiver is paused so the message isn't appended to while it's being processed. The `raw` list is created by extracting all the values from the `ir_in` object. These are normalised and decoded. Finally, if there's a message, it's printed out (you'll need to be connected to the REPL to see it), and the `ir_in` buffer is cleared and resumed for reading new messages.

Using these techniques, it's possible to create your own protocols or implement those of others. Morse code probably isn't a very good protocol to use in this respect, but has been useful as a teaching aid since it demonstrates the duration-based nature of infrared signals. While such infrared signals facilitate inter-device communication, you'd be hard pressed to describe the end result as a robust computer network. For that you'll need to use the radio capabilities of the micro:bit and ESP8266/32-based devices.

The micro:bit Radio

The radio hardware on the micro:bit makes available a custom wireless layer that is both simple to think about and uses very little memory and few resources. These features are exposed in MicroPython on the micro:bit as the `radio` module.

Conceptually, the `radio` module is very simple. Imagine you are the member of a children's gang, each of whom has a basic walkie-talkie. Everyone agrees beforehand the channel number to which their walkie-talkies are to be tuned. Once out and about, if you press the broadcast button and speak, everyone on the same channel will receive your message. Children intuitively understand this network topology simply because they use it so often themselves!

At a more technical level, messages are of a certain configurable length that can be up to 251 bytes long. The default length is 32 bytes. Incoming messages are put into a queue of configurable size. The default queue length is three, and the larger the queue, the more RAM is used. If the queue is full, new messages are ignored. Messages are broadcast at a certain power level in a range of zero (weakest) to 7 (strongest). The default value is 6 and more power means greater range, but this will use up batteries more quickly. The rate of throughput (i.e., speed of delivery) can be one of three pre-determined settings: 250,000 bits, 1 MB, or 2 MB a second. The default is 1 MB per second. Messages are broadcast and received on a preselected channel (numbered from zero to 100). In addition, messages can be filtered by address and group. The address is analogous to a house number, and group is like a named recipient at the specified address.

The radio API allows you to configure all of the parameters mentioned, as well as send and receive bytes. As a convenience for children, the radio module also makes it easy to send and receive strings. Use bytes to work with arbitrary data, although strings are remarkably flexible for many purposes.

Since the radio draws power, you have to explicitly turn it on to send or receive messages. If your application is in a fire-and-forget situation, this is handy since you only need to switch the radio on to send a message. Switch it off at all other times to conserve power. Since this is a child-friendly API, the two functions are called radio.on() and radio.off().

Use the radio.config(**kwargs) function to update the parameters, including the following:

length *(default, 32)*
Defines the maximum length of a message in bytes. The upper limit is 251.

queue *(default, 3)*
Specifies the number of message to store in the message queue.

channel *(default, 7)*
Defines the arbitrary channel to which the radio is tuned.

power *(default, 6)*
Indicates the strength of the broadcast signal.

address *(default, 0x75626974)*
An arbitrary name expressed as a 32-bit address used to filter incoming packets.

group *(default, 0)*
An 8-bit value (0-155) used with address when filtering messages.

data_rate *(default, 1 MB)*
Indicates the speed at which data throughput takes place.

If you find the radio has got into a bad state, you can call radio.reset() to return it to the sensible default settings.

Assuming the radio is on, to send a string simply use radio.send("Hello, World!"). Receiving a string is equally as simple: just do something like msg = radio.receive() to return the first message in the queue and to remove it from the queue to make way for subsequent messages. If the message queue is empty, then `radio.receive()` returns None.

Sending and receiving bytes is equally simple. Simply, `radio.send_bytes(message)` (where message contains bytes) and `message_bytes = radio.receive_bytes()`. Just like the string-based function, `radio.receive_bytes()` returns None if the message queue is empty; otherwise, it returns the first message in the queue and removes it to make space for new messages. If you are using buffers in your application, you also have the option to use `radio.receive_bytes_into(buffer)`. In this case, the next incoming message from the message queue is copied into the buffer object, trimming the end of the message if necessary. In order to help detect the state of the message queue, this function returns None if there are no pending messages or, if a message was received, an integer representing the length of the incoming message.

Let's put this together with a simple example:

```
from microbit import *
import radio

radio.config(channel=42)
radio.on()

while True:
    sleep(20)
    if button_a.was_pressed():
        radio.send("Hello")
    msg = radio.receive()
    if msg:
        display.scroll(msg, 80, wait=False)
```

The script begins by changing the radio configuration to channel 42 before powering up the radio. Next comes an infinite event loop. Each iteration of the event loop checks two things:

1. If button A was pressed, send the message "Hello".
2. If a new message is received, scroll it across the display.

Notice how close the description of the event loop is to the way the Python code is written. This pattern of sending messages when a certain event happens and checking for new messages at each iteration of the event loop is at the core of all applications that use the micro:bit's radio.

While this simple application is fun from an illustrative point of view, a far more useful application demonstrates how easy it is to create a complicated program that makes a collection of devices to work as a peer-to-peer mesh network.

Imagine you have a micro:bit stuck to the wall next to the light switch in every room of your house. They can be used as a house broadcast system (for example, to save you the effort of hollering up the stairs when dinner is ready). Each micro:bit is named after its location ("Kitchen", "Bathroom", "Lounge", and so on) and, depending on the room, has a list of pre-programmed messages to send from the specified room. I imagine you definitely want to have messages like "Yes" and "No" in the list of pre-programmed messages for all rooms, although "We've run out of toilet roll" only works if it's the bathroom-based micro:bit that has sent the message.

Pressing button "A" should enumerate the available messages (and the current message should be scrolled across the display to confirm the message has changed). Pressing button "B" should send the currently selected message and scroll the sent message on the sending micro:bit as a confirmation.

The message should also indicate the originating micro:bit so that everyone else can tell who sent the message.

Upon receiving a new message, the sender and content of the message should scroll across the display. Furthermore, because some micro:bits may be out of range of the originating device, the recipient device should immediately rebroadcast the message so everyone in the mesh of devices gets to hear about it. Of course, with all the devices rebroadcasting messages, there needs to be some way to mitigate the echo effect (where devices end up infinitely rebroadcasting the rebroadcasted messages). This is achieved with a simple message cache. Every message received becomes a key with an associated timestamp value. Messages in this cache are cleaned out after an arbitrary period of time. So, if the message exists in the cache, then it's been seen recently and should be ignored (thus stopping the infinite echo problem). This is a common pattern in peer-to-peer mesh networks where all members need to get a message and not every member of the network is connected, so devices automatically pass on messages to any other devices that are listening.

Here's one way to achieve this, as copiously annotated Python:

```python
import radio
from microbit import *

radio.config(length=64)
radio.on()

device_name = "Lounge"

messages = [
  "TV is free.",
  "Need logs for the fireplace.",
  "Fireplace needs cleaning.",
  "Lovely log fire!",
  "Nibbles and snacks available.",
  "Yes",
  "No",
]
```

```
message_cache = {}
cache_lifetime = 1000 * 5  # 5 seconds

position = 0

while True:
    sleep(20)
    # Sweep and clean the cache of stale messages.
    now = running_time()
    to_delete = []
    # Sweep.
    for key, timestamp in message_cache.items():
        # Check the age of the cached message.
        if now > timestamp + cache_lifetime:
            to_delete.append(key)
    # Clean the cache of out stale messages.
    for stale_message in to_delete:
        del message_cache[stale_message]
    # Cycle through the available messages.
    if button_a.was_pressed():
        position += 1
        # Skip back to the beginning if we reach the end.
        if position == len(messages):
            position = 0
        # Preview the newly selected message.
        display.scroll(messages[position], 50, wait=False)
    # Send the currently selected message.
    if button_b.was_pressed():
        # Message format is "sender:content".
        radio.send('{}:{}'.format(device_name, messages[position]))
    # Check for and display incoming message, rebroadcast if required.
    msg = radio.receive()
    if msg:
        if msg not in message_cache:
            # This is a new message, so store it in the cache.
            message_cache[msg] = running_time()
            # Rebroadcast it.
            radio.send(msg)
            # Display it in a friendly way.
            sender, message = msg.split(':')
            display.scroll('{} says: {}'.format(sender, message), 50,
                        wait=False)
```

After configuring the length of messages to be 64 bytes and switching on the radio, the script defines a name (to indicate its location) and a list of potential messages it could send. A clean cache for the messages is created, and a cache lifetime is defined in milliseconds. Setup completes with the initial message being that in position 0 of the message list (i.e., "TV is free.").

Next comes the event loop. This section is commented and demonstrates how the cache is maintained, how button events are handled, and how incoming messages are processed. It's all very simple and, given plenty of devices, rather fun.

Other devices will have a different version of the code. For example, the bathroom would revise the device_name and messages list to something along the lines of:

```
device_name = "Bathroom"

messages = [
  "Need more toilet roll.",
  "Run out of soap.",
  "Blocked drain.",
  "Missing toothbrush.",
  "The shower is free.",
  "Yes",
  "No",
]
```

I imagine the kitchen's version would look like this:

```
device_name = "Kitchen"

messages = [
  "Food is ready!",
  "Does anyone want a cup of tea?",
  "The dishwasher is finished.",
  "We need more milk.",
  "The table needs setting.",
  "Yes",
  "No",
]
```

Can you think of ways to improve the script? In the next chapter we will look at sound and music, so perhaps you could extend the script with speech or musical signals. Alternatively, a good rule of thumb is never to trust user-generated input (such as messages from other micro:bits). How might you protect your devices from badly formed messages or interference from other devices that are running a completely different application but which are broadcasting in the area?

Being able to coordinate many micro:bits is a lot of fun; and given the micro:bit's educational heritage, I also want to bring your attention to a simple application that works well in the classroom. It provides rather an interesting effect that beginner coders will enjoy: fireflies.

A *firefly* is a sort of bug that uses bioluminescence to signal to its friends. It's relatively simple to turn a group of micro:bits into fireflies if they all run this code:

```
import radio
import random
from microbit import display, Image, button_a, sleep

# Create the "flash" animation frames. Can you work out how it's done?
flash = [Image().invert()*(i/9) for i in range(9, -1, -1)]

radio.on()

while True:
    # Button A sends a "flash" message.
    if button_a.was_pressed():
        radio.send('flash')  # a-ha
    # Read any incoming messages.
    incoming = radio.receive()
    if incoming == 'flash':
        # If there's an incoming "flash" message display
        # the firefly flash animation after a random short
        # pause.
        sleep(random.randint(50, 350))
        display.show(flash, delay=100, wait=False)
        # Randomly re-broadcast the flash message after a
        # slight delay.
        if random.randint(0, 9) == 0:
            sleep(500)
            radio.send('flash')  # a-ha
```

Once again, the code contains comments to explain what is going on. Essentially, clicking button "A" on a micro:bit sends out a "flash" signal. Any recipient has a chance that it may rebroadcast the "flash" signal itself, thus propagating the signal over time. If a device receives a signal, it animates a flash on its display. The end result is something that looks like real-life fireflies signalling to each other. Given that a million of these devices have been handed out so that every 11-year-old in the UK has one, it's no stretch to imagine a class of 30 children enjoying such a spectacle. I also imagine this project would work well in a dark and dingy disco. Instead of pressing button "A", the "flash" message could be activated by a "shake" gesture if the device was worn on all the dancers' wrists. It would be quite a sight!

A final aspect of micro:bit radio is how to facilitate communication with non-micro:bit devices. The simple answer is you should use a listening microbit that's plugged into your computer. The code on this device simply listens for incoming messages and sends them to your computer via the USB serial connection (UART) and reads messages from your computer and broadcasts them to others in range. The following code demonstrates what I mean:

```
from microbit import *
import radio
```

```
radio.on()

while True:
    radio_msg = radio.receive_bytes()
    if radio_msg:
        uart.write(radio_msg)
    pc_msg = uart.read()
    if pc_msg:
        radio.send_bytes(pc_msg)
```

This is very simple code that should be easy to modify to your own ends.[4]

It is easy to read messages from USB serial in two ways: just connect to the device as if you were connecting to the REPL (the REPL uses USB serial to send and receive characters from your computer), or write a script so you can programmatically react to the incoming messages from the micro:bit. Here's an example script that uses the pySerial package to do just that:

```
"""
Listen to a connected micro:bit for incoming messages to which you can react
as needs apply.
"""
from serial.tools.list_ports import comports as list_serial_ports
from serial import Serial


def find_microbit():
    """
    Finds the port to which the device is connected.
    """
    ports = list_serial_ports()
    for port in ports:
        # Use the vendor and product ID to identify the micro:bit.
        if "VID:PID=0D28:0204" in port[2].upper():
            return port[0]
    return None


def get_serial():
    """
    Detect if a micro:bit is connected and return a serial object to talk to
    it.
    """
    port = find_microbit()
```

---

4 Sometimes you may find three "spare" bytes at the start of a radio message. These will always be, 1, 0, 1. They only occur if you mix send (that deals with strings) and receive_bytes (that deals with bytes). They are an artefact from making MicroPython compatible with a protocol from other radio-using platforms that target the micro:bit. Since we're only interested in the content of the message, it is safe to strip these out.

```
    if port is None:
        raise IOError('Could not find micro:bit.')
    return Serial(port, 115200, timeout=1, parity='N')


serial = get_serial()  # create the serial connection to the micro:bit


# Keep listening for bytes from the device. If any are received print them.
while True:
    msg = serial.read_all()  # Remember, msg will be bytes not a string.
    if msg:
        # At this point you could check the content of msg to react in more
        # complicated ways than just printing it. For example, you could use
        # serial.write(a_response) to re-broadcast a message from the
        # micro:bit.
        print(msg)
```

Using this method I have seen the micro:bit turned into a remote-control, game-pad-like device.[5]

The micro:bit's networking capabilities offer a lot of potential, especially in an educational setting. However, to be truly Internet of Things buzzword-compliant, we need to be able to connect to the internet. That means WiFi, and that means the ESP8266- and ESP32-based boards.

ESP8266/32 WiFi

 It's important to remember that the following code is for ESP8266 boards. The ESP32 port of MicroPython is still in active development, although these examples should work since there should be a consistent API between the two implementations.

By connecting a device to the internet, a new world opens up to you: the combined computing power of everything connected to the cloud, a plethora of useful APIs, and the ability to send and receive messages outside your current location. This is exciting: one of the magical attributes of "enchanted" objects is their ability to communicate with distant actors, services, and devices; a simple, classic example being a glowing orb that indicates the weather (the enclosed LEDs change colour depending upon the result of a remote call to a weather service for the weather forecast of some arbitrary area).

---

5 See, for example, Martin O'Hanlon's X-Wing in Minecraft controlled by a micro:bit (*https://www.youtube.com/watch?v=59KqWVwj_Cc*).

The first step is to get connected. There are two ways to do this: by connecting the board as a station (client) to an existing WiFi network or by using the board as an access point so other devices can connect to a WiFi network provided by your device. Both approaches are achieved via the network module as demonstrated in the following REPL interaction:

```
>>> import network
>>> station = network.WLAN(network.STA_IF)
>>> access_point = network.WLAN(network.AP_IF)
>>> station.active()
False
>>> station.active(True)
>>> station.scan()
[(b'WiFi Network Guest Access', b'61\xc4y\xfd?', 6, -58, 4, 0),
(b'Wifi Network', b'41\xc4y\xfd?', 6, -57, 4, 0)]
>>> access_point.active()
True
>>> access_point.ifconfig()
('192.168.4.1', '255.255.255.0', '192.168.4.1', '0.0.0.0')
>>> access_point.ifconfig(dns='208.67.222.222')
```

When instantiating a WLAN class, the network interface name should be specified. Two such objects were created in the preceding example: the station object with the STA_IF (station interface) name, and the access_point object with the AP_IF (access point interface) name.

Since the station doesn't know how to connect to any networks, it begins in an inactive state. After activation, it's possible to get a list of currently available networks expressed as tuples containing the SSID (network name), BSSID (the MAC address of the access point providing the network), channel, RSSI (signal strength), authentication mode, and hidden flag.[6]

Since the access_point provides a WiFi network to which others may connect, it starts in an active state. To learn what the IP-level network parameters are for the access point, call ifconfig. It returns the IP address, subnet mask, gateway, and DNS server. If you need to change any of these, just call the same method but pass in a named value (any of ip, subnet, gateway, or dns) as shown in the example.

When the device acts as an access point, it will show up with an SSID like MicroPython-121ce1. As a quick Google will tell you, the default password is micro pythoN (note the capital "N" at the end).

---

6 There are five possible values for authmode: 0 (open), 1 (WEP), 2 (WPA-PSK), 3 (WPA2-PSK), and 4 (WPA/WPA2-PSK). There are two modes for the hidden flag: 0 (visible) and 1 (hidden). For more information, see the relevant MicroPython documentation (*http://bit.ly/micropython-available-wireless*).

The first thing you should do is configure your WiFi so that you change the connection settings for your device if it is acting as an access point, or if you need your device to connect to a third-party WiFi.

To configure the device as an access point, you should use the `config` method as demonstrated in this continuation of the REPL session:

```
>>> access_point.config('essid')
'MicroPython-121ce1'
>>> access_point.config('authmode')
4
>>> access_point.config(essid='new_net_name', password='new_password')
>>> access_point.config('password')
Traceback (most recent call last):
  File "<stdin>", line 1, in <module>
  ValueError: unknown config param
```

The `config` method is used to query the value of a setting by passing in the setting name as a string. Alternatively, it is possible to update the setting by using it as a named argument (as when the network name and password were updated in the previous example). As the example shows, MicroPython won't reveal your password. It knows about the following parameters to define the settings for your access point network:

mac
    MAC address as bytes.

essid
    The WiFi access point's network name, as a string.

channel
    WiFi channel as integer.

hidden
    A boolean to indicate if the network is hidden.

authmode
    The supported mode of authentication as enumerated in a previous footnote.

password
    The password as a string.

 You should update the network name (essid) and password as soon as possible when you're running the device as an access point.

To configure your device to connect to an available WiFi network, ensure that it is active and then call the connect method:

```
>>> station.active(True)
>>> station.connect('Network Name', 'password123')
>>> station.ifconfig()
('192.168.178.190', '255.255.255.0', '192.168.178.1', '192.168.178.1')
```

Once connected, the ifconfig method tells you the device's IP address.

MicroPython is helpful by remembering such configurations for both station and access point modes of connection after you restart the device. In other words, you need only do this once. Upon restart, the device will attempt to resume or reconnect, given the status of the device when it was switched off. It is easy to disable the station or access point connections if you no longer need them:

```
>>> access_point.active(False)
```

Alternatively, you may just need to disconnect:

```
>>> access_point.disconnect()
```

If you ever need to check the status of the connections, you have two options. The isconnected method will return a boolean indication of the connection state; if the device is acting as a station, it will return True if the device is connected to a WiFi network and has an IP address; if the device is acting as an access point, it will return True if a station is connected to the device; otherwise, the result is False. Alternatively, the status method returns a list of possible states that the wireless connection could be in:

STAT_IDLE
No connection and no activity.

STAT_CONNECTING
Connecting in progress.

STAT_WRONG_PASSWORD
Failed due to incorrect password.

STAT_NO_AP_FOUND
Failed to connect because no access point replied.

STAT_CONNECT_FAIL
Failed due to other problems.

STAT_GOT_IP
Connection successful.

Once connected, it's time to have some fun over the internet. When the ESP8266 was first created, Damien attended a session of the London Python Code Dojo, a community group where "social coding" takes place, friends are made, and people learn from each other. He brought along some devices, gave a quick overview of the board, its capabilities, and how to connect to the network. All the attendees were encouraged to play. At the end, everyone came together to show what they'd managed to achieve. The example I'm going to use to demonstrate simple network connectivity was first hacked together by attendees at the dojo and later refined by Damien. Perhaps most importantly, all the code in the example will work with "regular" Python, thus demonstrating that MicroPython really is a comprehensive port of Python 3 in all its glory.

Python has a socket module with which one creates connections over the internet. MicroPython also has a version of this module, and we can use it to reach out to servers on the internet. Given a domain name, it's possible to get the associated IP address, create a socket, connect the socket to the remote IP address, and print out all the data received from the remote server.

The following example involves downloading data in a REPL session and assumes that you're connected to a WiFi network with internet access:

```
>>> import socket
>>> addr_info = socket.getaddrinfo("towel.blinkenlights.nl", 23)
>>> addr_info
[(2, 1, 0, '', ('94.142.241.111', 23))]
>>> server_addr = addr_info[0][-1]
>>> s = socket.socket()
>>> s.connect(server_addr)
>>> while True:
...     data = s.recv(500)
...     print(str(data, 'utf8'), end='')
...
```

Notice how the socket module's getaddrinfo method is used to turn a domain name into an IP address and how such information becomes the tuple server_addr. Armed with this information, a new socket is instantiated with a connection made using the server_addr. Finally, an infinite loop ensures that we receive 500 bytes of data at a time from the remote server and print this to the REPL session.

At this point you should be watching ASCII-mation *Star Wars* (*https://en.wikipe dia.org/wiki/ASCII_art#Animated_ASCII_art*) (Figure 10-1).

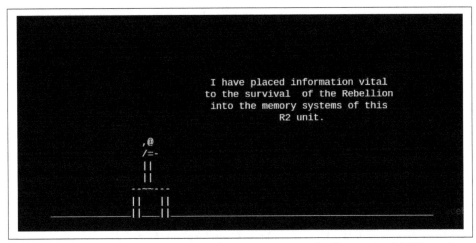

Figure 10-1. A screenshot of a MicroPython REPL session displaying the ASCII-mation version of Star Wars.

It is also possible to send data: create a socket as before, but use the `send` method to send bytes across the network. For example, here's how to read the home page of MicroPython via an HTTP GET request:

```
>>> import socket
>>> addr = socket.getaddrinfo('micropython.org', 80)[0][-1]
>>> s = socket.socket()
>>> s.connect(addr)
>>> s.send(b'GET / HTTP/1.1\r\nHost: micropython.org\r\n\r\n')
>>> data = s.recv(1000)
>>> s.close()
```

At this point the `data` object will contain the HTML (expressed as bytes) for the root of the website at micropython.org.

By now you are probably wishing that the popular `requests` module was available for MicroPython. It's advertised as, "HTTP for Humans" (*http://docs.python-requests.org/en/master/*) and makes it exceptionally easy to make web-based requests in Python code. Furthermore, since many of the use cases for the `requests` module are to API endpoints that respond to or reply with JSON data, it would be pretty helpful if there were a port of Python's `json` module, too.

It turns out that the `urequests` (micro requests) module for MicroPython mimics much of the core functionality of the popular `requests` module and is already built into the ESP8266 port of MicroPython, along with the standard `json` module. As a result, the previous seven lines of Python can be reduced to:

```
>>> import urequests as requests
>>> response = requests.get('http://micropython.org/')
```

Since most programatic interactions over HTTP are with endpoints that use JSON, it's relatively trivial to use fun services such as the Star Wars API hosted at *http://swapi.co/*:

```
>>> import urequests as requests
>>> response = requests.get('http://swapi.co/api/people/1/')
>>> dir(response)
['text', '__init__', '__qualname__', 'close', 'content', 'json', '__module__',
'encoding', 'raw', 'reason', '_cached', 'status_code']
>>> person = response.json()
>>> person['name']
'Luke Skywalker'
>>> person['homeworld']
'http://swapi.co/api/planets/1/'
>>> person['films']
['http://swapi.co/api/films/6/', 'http://swapi.co/api/films/3/',
'http://swapi.co/api/films/2/', 'http://swapi.co/api/films/1/',
'http://swapi.co/api/films/7/']
```

The `response` object has the expected `json` method to return a Python dictionary that can be interrogated in the usual way. Since the Star Wars API is read only, we'll use the JSON placeholder service to demonstrate sending data via HTTP:

```
>>> import urequests as requests
>>> import json
>>> data = json.dumps({'hello': 'world'})
>>> url = 'http://jsonplaceholder.typicode.com/posts'
>>> response = requests.post(url, data=data)
>>> response.json()
{'id': 101}
```

While it's fun to send messages and request data, it is also sometimes necessary to respond to incoming messages. While MicroPython has a version of Python's `asyncio` module, it's not currently a part of the ESP8266 port because of space and memory constraints. However, it is relatively simple to use Python's `socket` module to create a basic web server that returns a JSON representation of the digital state of all the device's GPIO pins:

```
import machine
import socket
import json

template = """HTTP/1.1 200 OK
Content-Type: application/json
Content-Length: {length}
Server: MicroPython

{json}"""

pins = [machine.Pin(i, machine.Pin.IN) for i in (0, 2, 4, 5, 12, 13, 14, 15)]
addr = socket.getaddrinfo('0.0.0.0', 80)[0][-1]
```

```
s = socket.socket()
s.bind(addr)
s.listen(1)

print('listening on', addr)

while True:
    cl, addr = s.accept()
    print('client connected from', addr)
    cl_file = cl.makefile('rwb', 0)
    while True:
        line = cl_file.readline()
        if not line or line == b'\r\n':
            break
    status = {str(p): p.value() for p in pins}
    data = json.dumps(status)
    response = template.format(length=len(data), json=data)
    cl.send(response)
    cl.close()
```

This example is an excellent template for other simple ESP8266-based servers. Apart from the pin-related lines, it's also just standard Python sockets programming. After setting up a socket to listen for incoming connections on port 80, the server enters an infinite loop that waits to process incoming connections. After reading the request until a blank line (it's at this point you may want to process the content of each line received to work out, for example, details of the request headers), the status of the pins is obtained, turned into a data string containing a JSON representation of the status, turned into a response by inserting the data into the HTTP response template, sent to the requestor, and ends by closing the client connection.

From a regular Python REPL, the client would see something like this:

```
>>> import requests
>>> esp8266_url = 'http://192.168.178.190/'
>>> response = requests.get(url)
>>> response
<Response [200]>
>>> from pprint import pprint
>>> pprint(response.json())
{'Pin(0)': 1,
 'Pin(12)': 1,
 'Pin(13)': 1,
 'Pin(14)': 1,
 'Pin(15)': 0,
 'Pin(2)': 1,
 'Pin(4)': 1,
 'Pin(5)': 1}
```

That should be familiar to many developers who have to consume common web-based APIs, endpoints, and web services.

Given MicroPython's embedded context, and the constrained nature of the devices upon which it runs, we will end this chapter with a quick look at the Message Queue Telemetry Transport (MQTT), a data transmission protocol common in the embedded world.

MQTT

The *Message Queue Telemetry Transport* (MQTT) is a lightweight publish/subscribe (pub/sub) messaging protocol. Devices (called clients) connect to a central server (called a broker) and subscribe to topics they are interested in. Clients may also publish messages to topics. Many clients may subscribe to a topic, and any messages published to the topic will be aggregated to the subscribers. The broker and MQTT protocol act as the common mechanism through which devices may connect and communicate. In this way, your ESP8266 device may communicate via a topic with any other MQTT-connected devices such as smart-home sensors and other Internet of Things devices connected to an MQTT broker.

MicroPython devices will act as clients in an MQTT network.

A popular cross-platform open source broker is Mosquitto (*http://mosquitto.org/*). I've seen it used effectively on a Raspberry Pi acting as a (well-hidden) broker for home automation projects.

Mosquitto is packaged for all major operating systems. You should consult the project's documentation for installation and configuration instructions.

There is no need to configure a topic; simply publishing a message to a topic will do. Topics use a naming convention similar to the path of a URL: there is a hierarchy with "/" (slash) used as the separator. It means topics can be organised under common themes, and naming conventions for new topics can evolve. For example, your devices could use the following naming convention for topic names concerning temperature measurements: `sensors/DEVICE_NAME/temperature`. The device would publish messages to a topic where the `DEVICE_NAME` is replaced with some unique identifier.

Clients subscribe to specific topics in order to receive any messages published to the topic. Alternatively, clients can use two wildcards (+ and #) to subscribe to all topics that match the wildcards.

The + wildcard is used to match any single level of a hierarchy. For example, `sensors/+/temperature` would specify a subscription to temperature readings on all devices, no matter what their name (hence the wildcard).

The # wildcard matches all the remaining levels of a hierarchy and therefore must always be used as the final character of a subscription specification. For example, a subscription to `sensors/+/#` would result in messages from all devices for all types of reading (not just `temperature`).

Just because a device is supposed to be connected to a network, it does not mean the network, connection, or device is working correctly. As a result, MQTT defines three levels of quality of service (QoS) that determine how hard a client or broker will try to ensure that a message gets through. Higher levels of QoS are more reliable, but have higher resource requirements:

- 0: the message is delivered just once with no confirmation.
- 1: the message is delivered at least once with a confirmation required.
- 2: the message is delivered just once but with a four-step handshake to ensure a reliable connection.

Messages can be sent and subscriptions made at any level of QoS. In other words, the client chooses the maximum level of QoS it will receive. For example, if client A publishes a message to a topic at QoS 2, then client B, who is subscribed to the topic at QoS 0, will only have messages delivered with QoS 0 despite the original message being published at QoS 2. Furthermore, if client A is subscribed to the topic at QoS 2, and client B publishes a message at QoS 0, then client A will only receive it at QoS 0. The QoS level of the message doesn't change to the QoS level of the subscription if the subscription's QoS is higher.

A message can be set to be retained by the broker. It will be kept even if all the currently connected subscribers have received it. However, if a new subscription is made to the topic of the retained message, then it will be sent to the client.

When connecting, a client sets a "clean session" or "clean start" flag. If set to false, then the session is set to be durable: if the client disconnects, any subscriptions will be retained, and subsequent QoS 1 or 2 messages will be stored until it manages to reconnect. If the "clean session" is true, all subscriptions will be removed when the client disconnects.

Finally, and rather morbidly, MQTT has the notion of a "will". When a client connects to a broker, it can send it a will. The *will* is a special message, with an arbitrary topic, QoS level, and retainment status just like any other message. It won't be sent unless the client unexpectedly disconnects from the broker.

The ESP8266 port of MicroPython contains a small and simple MQTT module called `umqtt` (micro MQTT). It supports most of the MQTT features, including publishing and subscription via a single client object. The handling of messages from subscribed topics is done by setting a callback.

To keep the code size small, only QoS 0 and 1 are supported in `umqtt`.

A connection with a broker is represented by an instance of the `MQTTClient` class. It provides all the methods needed to work with the broker. The following example demonstrates how to publish to a topic (in this case, recording when the device's "FLASH" button is pressed):

```
import time
import ubinascii
import machine
from umqtt.simple import MQTTClient


button = machine.Pin(0, machine.Pin.IN)

broker_address = '192.168.1.35'
client_id = 'esp8266_{}'.format(ubinascii.hexlify(machine.unique_id()))
topic = b'button'

client = MQTTClient(client_id, broker_address)
client.set_last_will(topic, b'dead')
client.connect()

while True:
    while True:
        if button.value() == 0:
            break
        time.sleep_ms(20)
    client.publish(topic, b'toggled')
    time.sleep_ms(200)

client.disconnect()
```

Things to note about this example are the way the `client_id` is created from a hexlified version of the machine's `unique_id`, the topic and message are represented as bytes, and a session is started and stopped by calls to `connect` and `disconnect`. It's possible to set the clean session with a `clean_session=False` argument for the `connect` method (it defaults to `True`).

Publication of a message on a topic is achieved by the `publish` method that takes the topic and message bytes. Additional optional arguments are `retain` (that defaults to `False`) and `qos` (that defaults to `0` but can be set to `1`—remember QoS 2 isn't supported for space saving reasons).

The `set_last_will` method has exactly the same signature as the `publish` method.

Subscription to a topic is also very easy:

```
import time
import ubinascii
import machine
from umqtt.simple import MQTTClient
```

```
def callback(topic, message):
    """
    Received messages are processed by this callback.
    """
    print((topic, message))


broker_address = '192.168.1.35'
client_id = 'esp8266_{}'.format(ubinascii.hexlify(machine.unique_id()))
topic = b'button'

client = MQTTClient(client_id, broker_address)
client.set_callback(callback)
client.connect()
client.subscribe(topic)

while True:
    client.wait_msg()
```

The script contains many similarities to that used for publishing. Differences include the definition of a `callback` function that does something with the topic and message it receives, the use of the `set_callback` method to connect the callback function to the client's subscriptions, the `subscribe` method to set a subscription to the specified topic, and the blocking `wait_msg` method that sits in a loop constantly polling for updates. This last method can be replaced with the nonblocking `check_msg` method. Use the latter if you have any foreground processing to do while the polling takes place. It's also possible to set a `qos` argument in the `subscribe` method (either 0 or 1).

CHAPTER 11

Sound and Music

Sound is an intriguing medium. We use it to signal (such as the ringing of a mobile phone), create art (with music) and communicate meaning (through speech). If a device can make sound, it can signal, make music, and maybe even speak to us. There's also something deeply satisfying in making things go "bloop".

Three of the devices are immediately able to create sound. The Circuit Playground Express has a built-in speaker, the PyBoard has an AMP audio skin and the micro:bit comes with modules for making sounds if you attach speakers to it via the GPIO pins.

This chapter explores all the ways in which you can make sounds, music, and speech with MicroPython.

Bleeps and Bloops

The speaker on the Circuit Playground Express, like all things that produce sound, needs to create vibrations in the air. The simplest way to do this is to switch the buzzer on and off very quickly, thus making it vibrate to create a sound whose pitch is determined by how quickly the vibrations are oscillating. The following code demonstrates how to make the buzzer go "bloop" for two seconds:

```
import audioio
import array
import time
import digitalio
from board import SPEAKER, SPEAKER_ENABLE

# Switch on the speaker for output.
speaker_enable = digitalio.DigitalInOut(SPEAKER_ENABLE)
speaker_enable.switch_to_output(value=True)

duration = 2
```

```
length = 8000 // 1760
wave = array.array("H", [0] * length)
wave[0] = int(2 ** 15 - 1)

with audioio.AudioOut(SPEAKER, wave) as speaker:
    speaker.play(loop=True)
    time.sleep(duration)
    speaker.stop()
```

By changing the `duration`, you change the length of time you hear the "bloop". Sound production happens in the final four lines where a `speaker` object is used to play a repeated waveform while the board sleeps for `duration` seconds after which the `speaker.stop` method makes it silent again.

Armed with such a simple source of "bloops", it's very easy to make something musically useful like a metronome (to which I've added NeoPixels to make sure the device flashes in time with the beat). Use the left and right buttons to change the tempo of the metronome, and press both to reset the device to the default tempo of 120 beats per minute (BPM):

```
import neopixel
import audioio
import digitalio
import array
import time
from board import *

np = neopixel.NeoPixel(NEOPIXEL, 10, auto_write=False)
left = digitalio.DigitalInOut(BUTTON_A)
left.pull = digitalio.Pull.DOWN
right = digitalio.DigitalInOut(BUTTON_B)
right.pull = digitalio.Pull.DOWN

length = 8000 // 1760
wave = array.array("H", [0] * length)
wave[0] = int(2 ** 16 - 1)

# Switch on the speaker for output.
speaker_enable = digitalio.DigitalInOut(SPEAKER_ENABLE)
speaker_enable.switch_to_output(value=True)

speaker = audioio.AudioOut(SPEAKER, wave)
bleep_duration = 0.02
default_tempo = 0.48
tempo = default_tempo
tempo_change = 0.02

while True:
    if left.value and right.value:
        tempo = default_tempo
    elif left.value:
```

```
        tempo = min(tempo + tempo_change, 2.98)
    elif right.value:
        tempo = max(tempo - tempo_change, 0.02)
    np.fill((0, 255, 0))
    np.write()
    speaker.play(loop=True)
    time.sleep(bleep_duration)
    speaker.stop()
    np.fill((0, 0, 0))
    np.write()
    time.sleep(tempo)
```

The code causes the device to simultaneously "bloop" and flash, then it sleeps for a duration whose value can be increased or decreased on button presses and repeats ad infinitum. The "bloop" can be extracted from the script with the following stand-alone bloop function (put this into a module called music.py on the Circuit Playground Express's file system):

```
import audioio
import digitalio
import array
import time
from board import *

# Switch on the speaker for output.
speaker_enable = digitalio.DigitalInOut(SPEAKER_ENABLE)
speaker_enable.switch_to_output(value=True)

def bloop(pitch, duration):
    length = 8000 // pitch
    wave = array.array("H", [0] * length)
    wave[0] = int(2 ** 16 - 1)
    with audioio.AudioOut(SPEAKER, wave) as speaker:
        speaker.play(loop=True)
        time.sleep(duration - 0.01)
        speaker.stop()
        time.sleep(0.01)  # add articulation silence
```

Notice that I've added a very short gap of silence for "articulation", that is to say, hearing the start of a note. If this didn't exist and we attempted to call bloop several times with the same pitch, we would just get a single continuous tone rather than several notes articulated by a very short moment of silence.

The resulting pitch of the note depends upon how fast the speaker oscillates between high and low values. This is a version of the PWM technique used with LEDs to change their brightness. A length is defined as the output sample rate divided by the desired pitch in Hz. A wave array of unsigned integers containing length items with a default value of 0 is created as a buffer. The first item in the buffer is set to the equivalent of "high". Since the speaker plays the buffer in a repeating loop, it means the membrane in the speaker vibrates (it's turned on and then off) once during each loop of the buffer, thus producing a period of the correct length of time to produce a frequency of the expected pitch. As a result, a continuous sounding tone is produced by the speaker.

As the pitch of the note changes, so does the length of the wave buffer, and hence the period and frequency. If the referenced pitch is lower, then the change from high to low happens less frequently because the loop is longer and thus takes more time to complete its period. As a result, the membrane of the speaker still rapidly vibrates, just not as frequently as for a higher pitch. A less frequent vibration lowers the audible pitch of the note produced. With this simple operation, we are able to make music, thanks to an ancient Greek philosopher and mathematician called Pythagoras.

While Pythagoras is well known for a certain theorem to do with right-angled triangles, he was also fascinated by finding mathematical patterns in nature, founded a sort of mathematical cult, and forbade his followers to eat meat and beans.

One of the mathematical patterns he was said to have discovered in nature is the relationships of sound to ratios. If you pluck a string of a certain length, it always produces the same pitch. Pythagoras noticed the pitch of a string was related to its length and the ratios of the lengths of the strings dictated how they related to each other. For example, if string A is twice the length of string B, it will make exactly the same note as B but the musical interval of a whole octave lower (the ratio 2:1). However, if the ratio between the two strings was 3:2, you'd get the musical interval of a perfect fifth. It is by building patterns from these ratios that we arrive at the scales of notes used in music. Most importantly for us, these ratios also apply to the values passed in as pitches into the bloop function.

Assuming the availability of the bloop function (as shown in the previous example) within a music module, we can experiment in the REPL to make new pitches:

```
>>> from music import bloop
>>> a = 440  # A in the middle of the treble clef
>>> bloop(a, 5)  # play A for 5 seconds
>>> bloop(a // 2, 5)  # play a low A
>>> bloop(a * 2, 5)  # play a high A
>>> bloop(int(a * 1.5), 5)  # play D (a perfect 5th lower)
>>> bloop(int(a * 0.66), 5)  # E (a perfect 5th higher)
```

By working our way through a musical pattern called the circle of fifths, we arrive at the frequencies needed to play all 12 notes in a chromatic scale. If I know the frequency value of the note "E", I can use it to work out the note a perfect fifth above it ("B"). If the "B" is in too high an octave, I simply double its value to get back to my original octave. By recursively working out what the next perfect fifth is and adjusting the octave, I get the following approximate values:[1]

```
notes = {
    'b': 493,
    'a#': 466,
    'a': 440,
    'g#': 415,
    'g': 392,
    'f#': 370,
    'f': 347,
    'e': 330,
    'd#': 311,
    'd': 294,
    'c#': 277,
    'c': 262,
}
```

Changing octaves is just a matter of doubling or halving these values as required. In fact, it is very easy to expand our work to include a simple musical domain-specific language (DSL), as the following script demonstrates:

```
import audioio
import digitalio
import array
import time
from board import SPEAKER, SPEAKER_ENABLE

# Switch on the speaker for output.
speaker_enable = digitalio.DigitalInOut(SPEAKER_ENABLE)
speaker_enable.switch_to_output(value=True)

notes = {
    'b': 493,
    'a#': 466,
    'a': 440,
    'g#': 415,
    'g': 392,
    'f#': 370,
    'f': 347,
    'e': 330,
    'd#': 311,
    'd': 294,
    'c#': 277,
    'c': 262,
}
```

1 I've had to round some numbers. The whole topic of accurately tuning using the circle of fifths is a complicated yet fascinating exploration of how music, maths, and physics are all interconnected.

```
def bloop(pitch, duration):
    length = 8000 // pitch
    wave = array.array("H", [0] * length)
    wave[0] = int(2 ** 16 - 1)
    with audioio.AudioOut(SPEAKER, wave) as speaker:
        speaker.play(loop=True)
        time.sleep(duration - 0.01)
        speaker.stop()
        time.sleep(0.01)  # add articulation silence


def play(tune):
    for note in tune:
        name, duration = note.split(':')
        bloop(notes[name], int(duration) / 8)

line1 = ['c:4', 'd:4', 'e:4', 'c:4']
line2 = ['e:4', 'f:4', 'g:8']
line3 = ['g:2', 'a:2', 'g:2', 'f:2', 'e:4', 'c:4']
line4 = ['c:4', 'g:4', 'c:8']
frere_jacques = line1 * 2 + line2 * 2 + line3 * 2 + line4 * 2

play(frere_jacques)
```

The play function expects a list of notes that follow the pattern note_name:duration.
The note name simply needs to be one of the keys in the notes dictionary that maps
to a pre-determined pitch value. The duration indicates the relative lengths of notes
to each other. The duration is divided by eight in the play function, so users are able
to use more convenient whole numbers instead of fractions of a second. At the end is
a musical example that causes the device to play Frére Jacques.

How would you improve this musical DSL? What features are missing? Think on this
while you read on, since it will be something to have in mind when we look at music
on the micro:bit later in this chapter.

Putting aside musical theory considerations, it's possible to change the waveform
used to produce notes so the quality of sound (its timbre) is changed. It's the equivalent of playing the same notes but with different musical instruments. Waveforms
with different shapes create different timbres. Four common waveforms are displayed
in Figure 11-1:

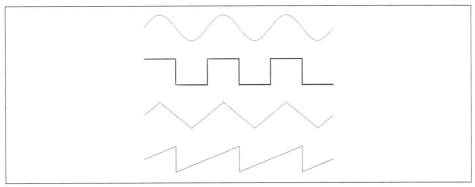

Figure 11-1. The shapes of four waveforms to change the timbre of the output. From the top: sine, square, triangle, and sawtooth.

Since each waveform has the same period (they all take the same time to complete a cycle), and it is the period that corresponds to the pitch, they all make the same note. However, because the values that change over the period of the waveform are different (thus making different shapes), they sound different.

The following example demonstrates how to make and play each type of waveform illustrated in Figure 11-1:

```
import audioio
import digitalio
import time
import array
import math
from board import SPEAKER, SPEAKER_ENABLE


# Switch on the speaker for output.
speaker_enable = digitalio.DigitalInOut(SPEAKER_ENABLE)
speaker_enable.switch_to_output(value=True)

length = 8000 // 440
sine = array.array("H", [0] * length)
triangle = array.array("H", [0] * length)
sawtooth = array.array("H", [0] * length)
square = array.array("H", [0] * length)


# The waveforms are created here.
for i in range(length):
    sine[i] = int(math.sin(math.pi * 2 * i / length) * (2 ** 15 - 1) + (2 ** 15))
    triangle[i] = abs(int(i * ((2 ** 15 - 1) // length)) - 2 ** 14)
    sawtooth[i] = int(i * ((2 ** 15 - 1) // length))
    if i < length // 2:
        square[i] = (2 ** 16 -1)
```

```
# Play each waveform.
print("Sine")
with audioio.AudioOut(SPEAKER, sine) as sample:
    sample.play(loop=True)
    time.sleep(2)
    sample.stop()

print("Triangle")
with audioio.AudioOut(SPEAKER, triangle) as sample:
    sample.play(loop=True)
    time.sleep(2)
    sample.stop()

print("Sawtooth")
with audioio.AudioOut(SPEAKER, sawtooth) as sample:
    sample.play(loop=True)
    time.sleep(2)
    sample.stop()

print("Square")
with audioio.AudioOut(SPEAKER, square) as sample:
    sample.play(loop=True)
    time.sleep(2)
    sample.stop()
```

The code speaks for itself: four buffers (one for each type of waveform) are created, populated with the expected values, and played one after the other. I'll leave it as an exercise for the reader to rewrite the Frére Jacques example but with different-sounding "instruments" based on the waveforms demonstrated.

It's not just programmed waveforms that can be played. Any audio data encoded as a mono, 16 kHz, 16-bit wav file can be played. In preparation for the next example, I've downloaded an MP3 of the Star Trek intercom "whistle" and converted it, using a piece of free software called Audacity (*http://www.audacityteam.org/*), to the format described. Next, I copied it onto the file system of the Circuit Playground Express as the sound.wav file:

```
import board
import audioio
import digitalio
from board import SPEAKER, SPEAKER_ENABLE

# Required for CircuitPlayground Express
speaker_enable = digitalio.DigitalInOut(SPEAKER_ENABLE)
speaker_enable.switch_to_output(value=True)

f = open("sound.wav", "rb")
speaker = audioio.AudioOut(SPEAKER, f)

speaker.play()
```

```
while speaker.playing:
    pass  # Block
```

Upon startup, the device whistles in that well-known way. Those of you familiar with Python will realise the code is merely opening the sound.wav file and referencing it as the source stream when the AudoOut class is instantiated. That's all there is to it!

Before we get onto musical micro:bit matters, we should turn our attention to the PyBoard and its AMP audio skin. This affords us an opportunity to explore yet another way to make interesting sounds.

The AMP skin has a small built-in speaker that's connected to a digital-to-analog converter (DAC) via a small amplifier. The DAC takes a digital value and turns it into an analog value represented by a voltage between 0 and 3.3 volts.

Playing a sound involves writing to the DAC in a way that is conceptually similar to that of the Circuit Python Express, the classic example being a sine wave at a certain pitch measured in Hertz:

```
import math
from pyb import DAC


buf = bytearray(100)
for i in range(len(buf)):
    buf[i] = 128 + int(127 * math.sin(2 * math.pi * i / len(buf)))

pitch = 440  # concert A
dac = DAC(1)
dac.write_timed(buf, pitch * len(buf), mode=DAC.CIRCULAR)
```

By changing the waveform, you get a different timbre (quality of sound), but you may notice it's too quiet. The volume of the amplifier can be controlled as an I²C device:

```
import pyb


def volume(volume):
    pyb.I2C(1, pyb.I2C.MASTER).mem_write(volume, 46, 0)
```

Valid values for volume range from 0 (quietest) to 127 (loudest), so you should experiment to find a comfortable level.

Can you work out how to get the PyBoard to play tunes? It's conceptually very similar to the Circuit Playground Express and left as an exercise for the reader.

An eccentric solution is to bit bang numbers to the DAC and listen. Here's how to re-create the white noise of a mistuned radio:

```python
import pyb
import random
from pyb import DAC


def volume(val):
    pyb.I2C(1, pyb.I2C.MASTER).mem_write(val, 46, 0)

volume(127)
dac = DAC(1)
while True:
    dac.write(random.randint(0, 256))
```

This isn't very pleasant to listen to, but it proves a point: it is relatively easy to control the sound with just a stream of numbers (in this case, a random stream of numbers). What would happen if the stream of numbers contained patterns? Could we be like Pythagoras and discover patterns in numbers that make interesting sounds?

Of course we can! Check this out for size:

```python
import pyb
from pyb import DAC


def volume(val):
    pyb.I2C(1, pyb.I2C.MASTER).mem_write(val, 46, 0)

volume(127)
dac = DAC(1)
t = 0
while True:
    dac.write(int(t*((t>>9|t>>13)&25&t>>6)) % 256)
    #dac.write(int(t*((15&t>>11)%12)&55-(t>>5|t>>12)|t*(t>>10)*32) % 256)
    #dac.write(int((t*9&t>>4|t*5&t>>7|t*3&t//1024)-1))
    t += 1
```

This script isn't much different from the white noise example, but it has a radically different output. There are three formulae that take t (representing time) and generate interesting patterns of numbers such that musical effects including, bleeps, bloops, beats, bass lines, and fragments of melody emerge from the speaker.[2] Each formula creates a different musical effect. If you're interested in how these musical patterns are created from such seemingly short fragments of code, you should start with the CounterComplex (*http://bit.ly/algorithmic-symphonies*) blog posts that investigate the phenomenon.

---

2 Two of the formula are commented out, so only the first one is played.

At this point, I believe Beethoven nailed it when he said:

> *O Freunde, nicht diese Töne! Sondern laßt uns angenehmere anstimmen, und freudenvollere!*[3]
> —Ludwig van Beethoven, *Ninth Symphony in D minor*

Music

MicroPython on the micro:bit includes a powerful `music` module that comes with an easy-to-learn musical DSL. It also comes with lots of tunes already built into the device. These tunes were created for two reasons:

1. By providing tunes for kids, we make it easy for them to use music in their games, programs, and projects. We included tunes that would work well as signals or represent emotions and events.
2. By looking at the built-in melodies, beginner programmers can work out how a particular musical effect is made.

Since the device doesn't include a speaker, you'll need to connect it as shown in Figure 11-2.

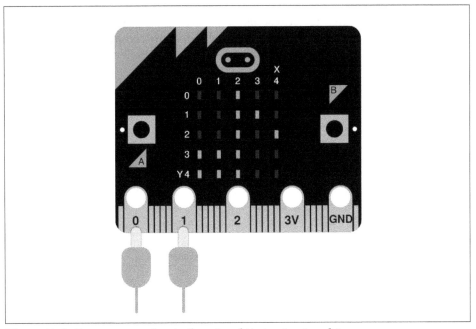

Figure 11-2. Connect a speaker to the micro:bit via pins 0 and 1

---

3 Oh friends, not these sounds! Let us instead strike up more pleasing and more joyful ones!

To check that it is working, the following REPL example will play a concert "A" at 440 hertz for 10 seconds. If you don't hear any sound, check the connections between the micro:bit and the speaker (and if the speaker is powered, make sure it's on!):

```
>>> import music
>>> music.pitch(440, 10000)
```

Assuming you get sound, it's very simple to turn the device into a musical instrument with the following script:

```
import music
from microbit import accelerometer


while True:
    music.pitch(accelerometer.get_x(), 20)
```

As you move, the device along the x-axis, the pitch of the sound will change, giving the effect of a strangled cat. The important point is that sound is a great way to provide feedback about change, in this case, the reading from the accelerometer. One way to improve the "instrument" and make it sound less like a strangled cat and more like something playing actual notes, is to quantize the accelerometer readings. Quantization is when a range of possible values is mapped to a single value. In the following example, if the accelerometer readings move around within certain ranges, only a single, constant pitch will be played. Instead of sliding around in pitch, quantization ensures that the changes only skip to notes in a certain pre-defined scale:

```
import music
from microbit import accelerometer

buckets = [
    262,  # C
    294,  # D
    330,  # E
    392,  # G
    440,  # A
]

while True:
    reading = abs(accelerometer.get_x())
    bucket = min(4, max(0, reading // 200))  # quantize!
    music.pitch(buckets[bucket], 20)
```

A pentatonic (five note) scale is defined as a list of pitches, each of which represents a bucket containing a range of possible accelerometer values. The accelerometer is read as an absolute (nonnegative) value. This determines the position within the buckets list so that a suitable pitch can be selected. Since the range of the accelerometer readings is approximately 0 to 1,000, the bucket is a value clamped between 0 and 4 (the available positions in the buckets list). The end result is a micro:bit that plays folk-like pentatonic melodies rather than sounding like a strangled cat.

Rather than trying to perform melodies with the micro:bit as an instrument, it is possible to play the built-in melodies like this:

```
import music
```

```
music.play(music.NYAN)
```

The `music` module contains a number of capitalised constants that define melodies with a simple and easy-to-use musical, domain-specific language. The names of these built-in melodies are descriptive of their musical content: BADDY, BA_DING, BIRTHDAY, BLUES, CHASE, DADADADUM, ENTERTAINER, FUNERAL, FUNK, JUMP_DOWN, JUMP_UP, NYAN, ODE, POWER_DOWN, POWER_UP, PRELUDE, PUNCHLINE, PYTHON, RINGTONE, WAWAWAWAA, WEDDING.

If you evaluate a built-in melody, you'll discover it's just a tuple of notes:

```
>>> import music
>>> music.DADADADUM
('r4:2', 'g', 'g', 'g', 'eb:8', 'r:2', 'f', 'f', 'f', 'd:8')
```

The musical DSL makes it easy to make new melodies since a tune is merely an ordered collection of notes. Each note has a note name (such as C, F#, or Bb), an indication of the octave containing the note, and a duration. Octaves are numbers between 0 (lowest) to 8 (highest), with middle C being found in octave 4 (the default octave if none is first given). Note names are case insensitive, so "a" is the same as "A". Durations are also numbers: the higher the value, the longer the note will sound. Durations are related to each other; for instance, a duration of 4 will last twice as long as a duration of 2. If you use the note name, R then MicroPython will play a silence for the specified duration.

Notes are written as a string of characters like this: NOTE_NAME[octave][:duration]. For example, "C1:6" refers to the note named "C" in octave number 1 to be played for a duration of 6. If this feels familiar, it's because the very simple musical DSL created in the Circuit Playground Express example works in a similar manner.

To create a melody, just make a list containing notes, as defined in the manner described.

MicroPython helps you to simplify melodies. It will remember the octave and duration values until you change them. As you can see from the DADADADUM example, not all the notes have the octave or duration annotated on them; they re-use the previously set values for these attributes. The octave and duration values only change when they need to, making the melody easier to read and quicker to type, as demonstrated in the following tune:

```
tune = ["C4:4", "D", "E", "C", "C", "D", "E", "C",
        "E", "F", "G:8", "E:4", "F", "G:8",
```

```
        "G:2", "A", "G", "F", "E:4", "C", "G:2", "A", "G", "F", "E:4", "C",
        "C", "G3", "C4:8", "C:4", "G3", "C4:8"]
```

Playing a melody means calling the play method. As demonstrated earlier, it expects a list of note definitions that define the melody to play, but can also receive other optional arguments such as wait, which, if set to False, makes the call nonblocking; and loop, which if set to True, repeats the melody until stop is called (see below) or the blocking call is interrupted. These should feel familiar, since they work in exactly the same way, as the show method associated with the display object. The following example will play the tune defined in the last example in a nonblocking manner and will keep repeating it:

```
music.play(tune, wait=False, loop=True)
```

The pitch method, used to play notes of a specific frequency, can also become nonblocking in the same way. As mentioned, the stop method stops all music playback (so it would silence the continuously looping tune started in the previous example).

It is possible to change the tempo of musical playback with the set_tempo method. It takes two arguments, the number of ticks that constitute a beat, and the number of beats per minute (BPM). A *tick* is what you specify as the duration of a note in the musical DSL. This method has default values for both arguments so you could, for example, just change the tempo with: set_tempo(bpm=180). A get_tempo method returns the currently set tempo as a tuple representing the ticks and BPM.

If anything goes wrong, use the reset method to put everything back to its default state (ticks, bpm, duration, and octave).

Speech

A computer that can play music is interesting, but a computer that can talk feels more "human".

If your only graphical user interface (GUI) is a 5 x 5 LED display, then it's hard to convey information to the user. Making the device talk is one way to give information in a medium that is fun, efficient, and useful. It is for this reason that a very simple software speech synthesiser was built into MicroPython on the micro:bit. Given the anthropomorphic characteristics of the physical design of the hardware (it looks like a face), the addition of a speech sythesiser just adds to its charm.

Connect a speaker to the device, import the speech module, and use the say function to make the micro:bit talk:

```
import speech


speech.say("Hello, World!")
```

It is a remarkably flexible speech synthesiser, since it's possible to change various characteristics of the voice:

Pitch
> How high or low the voice sounds (0 = high, 255 = low)

Speed
> The quickness of delivery (0 = fast, 255 = slow)

Mouth
> How tight-lipped or overtly enunciated the voice sounds (0 = tight lipped, 255 = overly enunciated)

Throat
> How tense or relaxed is the tone of voice (0 = tense, 255 = relaxed)

There doesn't appear to be any accepted way to work out how a voice's settings will sound; it's just a matter of experimentation. For example, the following is particularaly DALEK like in timbre:

```
speech.say("I am a DALEK - EXTERMINATE", speed=120, pitch=100, throat=100,
           mouth=200)
```

The `say` function also understands four punctuation marks: hyphen ("-") inserting a short pause, comma (",") inserting a pause double the length of the hyphen, full stop (".") and question mark ("?") end sentences with a long pause. The full stop causes the pitch to fall, whereas the question mark causes it to rise.

The `say` function is convenient. It makes it easy to write English and produce speech. Unfortunately, it's not always accurate. To ensure an accurate delivery of the expected speech, you should use *phonemes*, the smallest perceptually distinct units of sound that distinguish different words. They are the building blocks of speech.

The `pronounce` function takes a string containing a simplified version of the International Phonetic Alphabet and optional annotations to indicate inflection and emphasis. This has the advantage of not having to know how to spell; you only need to know how to say the word in order to spell it phonetically. All the phonemes understood by the synthesizer are listed with examples of their common usage in parentheses:

Simple vowels			
IY	f(ee)t	OH	c(o)ne
IH	p(i)n	UH	b(oo)k
EH	b(e)g	UX	l(oo)t
AE	S(a)m	ER	b(ir)d
AA	p(o)t	AX	gall(o)n
AH	b(u)dget	IX	dig(i)t
AO	t(al)k		

Diphthongs	
EY	m(a)de
AY	h(igh)
OY	b(oy)
AW	h(ow)
OW	sl(ow)
UW	cr(ew)

Voiced consonants			
R	(r)ed	D	(d)og
L	a(ll)ow	G	a(g)ain
W	a(w)ay	J	(j)u(dg)e
W	(wh)ale	Z	(z)oo
Y	(y)ou	ZH	plea(s)ure
M	Sa(m)	V	se(v)en
N	ma(n)	DH	(th)en
NX	so(ng)		
B	(b)ad		

Unvoiced consonants	
S	(S)am
SH	fi(sh)
F	(f)ish
TH	(th)in
P	(p)oke
T	(t)alk
K	(c)ake
CH	spee(ch)
/H	a(h)ead

Nonstandard phonemes	
YX	diphthong ending (weaker version of Y)
WX	diphthong ending (weaker version of W)
RX	R after a vowel (smooth version of R)
LX	L after a vowel (smooth version of L)
/X	H before a non-front vowel or consonant - as in (wh)o
DX	T as in pi(t)y (weaker version of T)

Special phonemes	
UL	sett(le)
UM	astron(om)y
UN	functi(on)
Q	kitt-en (glottal stop)

Pass in phonemes as a string like this:

```
speech.pronounce("/HEHLOW")  # "Hello"
```

If you pass invalid phonemes, a `ValueError` exception is raised.

Phonemes are classified into two groups: vowels and consonants.

Vowels are further categorised as simple vowels, which don't change their sound as you say them, or diphthongs, which start with one sound and end with another. For example, the word "oil" contains a diphthong: the "oi" starts with an "oh" sound but changes to an "ee".

Consonants are also subdivided into two groups: voiced, which require the speaker to use their vocal chords to make a sound (such as "L", "N", and "Z"), and unvoiced, which are produced by rushing air (such as "P", "T", and "SH").

Sometimes spelling with phonemes is counterintuitive. For example, the word "adventure" has a "CH" in it. The rule of thumb is to think about how the words sound, not how you would spell them. It is recommended that you experiment until you get the desired effect.

If you are stuck for where to start, you should use the result of the `translate` method. It tells you how the speech synthesiser would have turned plain English into phonemes:

```
>>> speech.translate("Hello")
' /HEHLOW'
```

To make the speech sound more natural and understandable, there is a built-in stress system to add inflection or emphasis. It consists of eight stress markers indicated by the numerals 1–8. Such markers should be inserted immediately after the vowel to be stressed. The expression "/HEHLOW" is rather robotic and can be made friendlier with a stress marker, "HEH3LOW".

The stress system works by raising or lowering pitch and elongating the associated vowel sound depending on the number you give:

1. Very emotional stress
2. Very emphatic stress
3. Rather long stress
4. Ordinary stress
5. Tight stress
6. Neutral (no pitch change) stress
7. Pitch-dropping stress
8. Extreme pitch-dropping stress

Such stress markers help pronounce difficult words correctly. For example, if a syllable is not enunciated sufficiently, use a neutral stress marker. It is also possible to elongate words:

```
speech.pronounce("/HEH5EH4EH3EH2EH2EH3EH4EH5EHLP.")
```

Finally, and rather remarkably, it is possible to make MicroPython sing by annotating a pitch related number onto phonemes. The lower the number, the higher the pitch, with numbers roughly translating into musical notes, as shown in Figure 11-3:

Figure 11-3. Pitch number for singing notes

Pitch annotations are a pre-pended hash ('#') sign followed by the pitch number and then the phoneme. The pitch will remain the same until a new annotation is given. Such annotations are only understood if you use the `sing` function like this:

```
import speech

solfa = [
    "#115DOWWWWWW",   # Doh
    "#103REYYYYYY",   # Re
    "#94MIYYYYYY",    # Mi
    "#88FAOAOAOAOR",  # Fa
    "#78SOHWWWWW",    # Soh
    "#70LAOAOAOAOR",  # La
    "#62TIYYYYYY",    # Ti
    "#58DOWWWWW",     # Doh
]
song = ''.join(solfa)
speech.sing(song, speed=100)
```

To extend the note in duration, use a repeated vowel or voiced consonant phonemes (as demonstrated in the preceding code example). Diphthongs are extended by breaking them into their component parts. For example, "OY" can be lengthened as "OHOHIYIYIY".

What can be achieved by combining speech synthesis and music?

In 1961, the very first singing computer, an IBM 7094, performed a version of "Daisy Bell". This, in turn, inspired Stanley Kubrick to use the song in his 1968 movie, *2001: A Space Odyssey*. The ship's rogue computer, HAL 9000, sings the refrain as an astronaut switches it off at the end of the film. Obviously, "Daisy Bell" is a historic piece when it comes to singing computers, and it is in this tradition that the following code demonstrates how to make a micro:bit sing a song:

```
import speech

line1 = [
    '#26DEYYYYYYYYYY',
    '#31ZIYIYIYIYIYIYIYIY',
    '#39DEYYYYYYYYYY',
    '#52ZIYIYIYIYIYIYIYIY',
    '#46GIXV',
    '#42MIYIY',
    '#39YAOW',
    '#46AEAEAEN',
    '#39SERER',
    '#52DUXUXUXUXUXUXUXUXUXUXUXUX' ]

line2 = [
    '#35AYYYYYYMM',
    '#26/HAEAEAEAEAEAEF',
    '#31KREYYYYYYY',
```

```
        '#39ZIYIYIYIYIYIYIYIY',
        '#46AXLL',
        '#42FAOR',
        '#39DHER',
        '#35LUHUHUHV',
        '#31AXAXV',
        '#35YUXUXUXUXUXUXUXUXUXUX']

    line3 = [
        '#31IHT',
        '#29WOWNT',
        '#31BIY',
        '#35ER',
        '#26STAYYYYY',
        '#31LIHSH',
        '#35MAE',
        '#39RIXIXIXIXIXIXIXIXIXIXIXIXIXIXJ',
        '#35AYY',
        '#31KAEAEAEAENT',
        '#39ER',
        '#46FAOAOAORD',
        '#39ER',
        '#46KAA',
        '#52RIXIXIXIXIXIXIXIXIXIXIXIXIXIXJ']

    line4 = [
        '#52BUHT',
        '#39YUXUXL',
        '#31LUXK',
        '#35SWIYIYIYIYT',
        '#52ER',
        '#39PAAAAAAN',
        '#31ER',
        '#35SIYIYIYT',
        '#31UHV',
        '#29ER',
        '#26BAY',
        '#31SIH',
        '#39KUXL',
        '#35MEYYYYD',
        '#52FER',
        '#39TUXUXUXUXUXUXUXUXUXUXUX']

    speech.sing(''.join(line1))
    speech.sing(''.join(line2))
    speech.sing(''.join(line3))
    speech.sing(''.join(line4))
```

For ease of reading, I've split each line into its own list, containing strings representing each vowel/note combination. At the very end, the lines are sung one after the other by joining each combination into a single string per line. The best strategy for judging the durations of notes is trial and error.

Finally, both the pronounce and sing functions can change timbre in the same way as the say function, using the speed, mouth, and throat named arguments. Once again, experimentation is the best strategy. Why not change the way "Daisy Bell" is sung?

Armed with the information contained in this chapter, you will be able to make your devices bleep, bloop, make music, talk, and sing for the entertainment and education of your users. It is in this spirit that I'll end this chapter with a fun script to make a beat-boxing micro:bit:

```python
import speech
from microbit import sleep, button_a, button_b, display, Image

gap = 220  # How long a silence should be.
bass_drum = "BUH"  # Sound of a beat box bass drum
snare = "CHIXIX"  # Sound of a beat box snare
roll = "DGDG"  # Sound of a beat box drum roll
rest = ""  # Represents a rest of "gap" duration

# Two sequences (lists) of beats. One beat per line.
beats1 = [  # Mellow
    bass_drum, rest, rest, rest,
    snare, rest, rest, rest,
    bass_drum, bass_drum, bass_drum, rest,
    snare, rest, roll, roll,
]

beats2 = [  # Hardcore
    bass_drum, snare, snare, snare,
    bass_drum, snare, roll, roll,
    bass_drum, bass_drum, bass_drum, snare,
    bass_drum, roll, roll, bass_drum,
]

# Play a sound or silence.
def beat_box(sound):
    if sound:
        display.show(Image.HEART)
        sleep(10)
        display.clear()
        speech.pronounce(sound)
    else:
        sleep(gap)

# Play all sounds in "beats" sequence.
def play(beats):
    for beat in beats:
        beat_box(beat)

selected = beats1  # Default beat sequence
# Keep on looping over the selected sequence
while True:
```

```
# Change sequence with buttons A and B
if button_a.was_pressed():
    selected = beats1
elif button_b.was_pressed():
    selected = beats2
# Finally play the selected sequence
play(selected)
```

## CHAPTER 12

Robots

Robots are cute in an "all humans must die" sort of a way; and, thanks to the educational uses of devices like the micro:bit, many makers have created robots that demonstrate how robotics can be a relatively simple and fun endeavour. This chapter explores two robotics projects that show how to use MicroPython to make your very own robotic invasion. Both use the micro:bit, although the techniques discussed are easy to transfer to other boards running MicroPython.

Trundle Bot

This bot trundles around on wheels. It has an analog distance sensor on the front to detect objects in its way. If something blocks its way, it rotates left or right until no blockage is detected, then it continues on its way. It's a very simple bot that can be made in about an hour with only a few parts. The code to drive the bot is also beautifully simple and demonstrates how to drive servo motors to give the bot movement.[1]

The bot was successfully used in a bot-building workshop at EuroPython 2016 and, because of its simplicity, was easy to modify and change to suit the available building materials and aims of the builders (who included experienced Python programmers, their nontechnical partners, and children).

The minimum parts required are inexpensive:

- A micro:bit
- 2 9g continuous rotation servos

---

[1] This bot was designed by the exceptionally talented Radomir Dopieralski. Radomir was one of the many volunteers who helped bring MicroPython to the micro:bit. His passion is making robots for MicroPython boards, and you can find many examples of his work on hackaday.io.

- A Pololu Carrier with Sharp GP2Y0A60SZLF Analog Distance Sensor, 3 V
- 2 wheels
- A Pololu caster ball
- A portable power source to provide between 3.3 V and 4.2 V

There also needs to be some means of making a chassis and wires to connect the components together. Double-sided sticky tape, rubber bands, and googly eyes are also helpful for assembling the device. Don't be too worried about making something that looks as well constructed as Figure 12-1. The point of this robot is to make something that just works. Once it's working, you can improve the construction. For example, Figures 12-2 and 12-3 show a homemade version of the bot made out of cardboard, sticky-backed plastic, rubber bands, and googly eyes. I'm sure you'll agree it has a certain charm about it (and more importantly, it's something a beginner interested in robotics would have fun constructing).

Figure 12-1. A very simple trundlebot

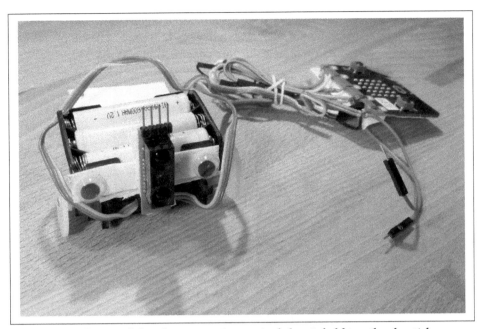

Figure 12-2. The trundlebot showing an improvised chassis held together by sticky-backed plastic, twisted wires, tape, and a rubber band

Figure 12-3. The assembled trundlebot (the micro:bit has sticky backed plastic to hold it in place). Googly eyes make it friendly.

The trick is to use your imagination and have fun. Most importantly, for the robot to work, it needs to be wired up correctly. While Figure 12-4 may at first look complicated, you will soon realise how simple the bot's construction really is.

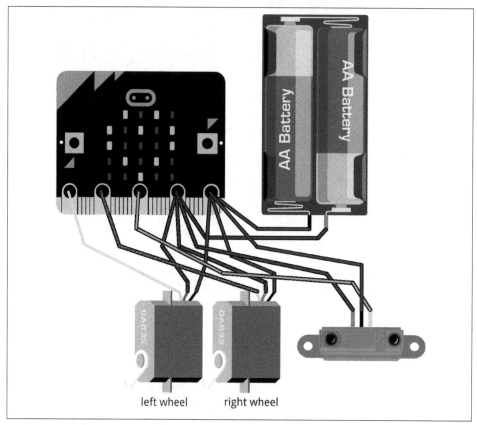

Figure 12-4. The trundlebot wiring diagram

Assuming you have connected the bot and assembled some sort of chassis, the next task is to drive the servo motors to make it move and take readings from the distance sensor so it won't bump into things.

The servos are physically connected to pins 0 and 1, through which analog signals are sent to control the direction and speed of rotation. The pulse width of the signal is the attribute that enables control. A pulse width of some arbitrary duration corresponds to a stopped motor. A pulse width either smaller or larger than the stopped value causes rotation in one direction or the other. The further away from the stop value the pulse width becomes, the greater the speed or rotation.

This functionality is wrapped up in a Servo class:

```python
import microbit

class Servo:
    def __init__(self, pin, trim=0):
        self.pin = pin
        self.trim = trim
        self.speed = 0
        self.pin.set_analog_period(20)

    def set_speed(self, speed):
        self.pin.write_analog(int(25 + 100 * (90 + speed) / 180 + self.trim))
        self.speed = speed
```

The class is initialised with a reference to the physical pin used to drive the motor and an argument called trim. *Trimming* is simply making fine adjustments to something; and, since servo motors are notoriously inconsistent in performance characteristics, there needs to be some way to adjust them. The speed attribute represents how fast the servo is moving, and the configuration is completed by setting the period (frequency) for PWM with the set_analog_period method of the pin. The set_speed method changes the speed of the servo. This can range from -90 (full backwards), via 0 (stop) to 90 (full forwards).

The next piece of the puzzle is representing the robot itself:

```python
class Robot:
    def __init__(self):
        # Remember to check the trim values.
        self.left_servo = Servo(microbit.pin0, 2)
        self.right_servo = Servo(microbit.pin1, 1)

    def go(self, distance):
        microbit.display.show(microbit.Image.ARROW_S)
        self.left_servo.set_speed(-90)
        self.right_servo.set_speed(90)
        microbit.sleep(int(distance * 2000 / 17))
        self.stop()

    def turn(self, angle):
        if angle > 0:
            microbit.display.show(microbit.Image.ARROW_E)
            self.left_servo.set_speed(-90)
            self.right_servo.set_speed(-90)
            microbit.sleep(int(angle * 64 / 9))
        else:
            microbit.display.show(microbit.Image.ARROW_W)
            self.left_servo.set_speed(90)
            self.right_servo.set_speed(90)
            microbit.sleep(int(-angle * 64 / 9))
        self.stop()
```

```
def stop(self):
    microbit.display.show(microbit.Image.DIAMOND)
    self.left_servo.set_speed(0)
    self.right_servo.set_speed(0)

def get_distance(self):
    return microbit.pin2.read_analog()
```

This allows us to pull all the components together under the control of a single instance of the Robot class.

Initialisation involves creating two new instances of the Servo class: one for each servo motor. As the comment says, if the robot veers off course when attempting to drive in a straight line, this is where to adjust the trim. The other methods provide all the functionality you need: go, turn, stop and get_distance. The go method uses its distance argument to work out how long to wait while the servo motors are active. Notice how the left and right motors work in opposite directions since they're mirrored in how they are arranged on the physical device (and thus the direction of travel is reversed). The turn method works in a similar way but given an angle argument, and the stop method does exactly what it says. All three of these methods work by actuating the servo motors in a meaningful way in order to hide the implementation details (at this level of abstraction in the code, we just want developers to think in terms of the robot rather than servo motors). In this spirit, the get_distance method simply returns the analog reading from the distance sensor connected to pin 2. This is an inverse reading: smaller means further away.

Making the robot work requires instantiating the Robot class and driving it from inside a simple event loop:

```
robot = Robot()
while True:
    robot.go(5)
    if robot.get_distance() > 700:
        robot.turn(20)
        left_distance = robot.get_distance()
        robot.turn(-40)
        right_distance = robot.get_distance()
        robot.turn(20)
        if left_distance < right_distance:
            robot.turn(60)
        else:
            robot.turn(-60)
```

Whilst in the event loop, the robot is made to go forward. However, if the robot gets a high result from the get_distance call, it follows a very simple algorithm: measure the distance of things to both the right and left of the current position, check which reading has the most available distance to move forward, and then turn in that direction. While this isn't HAL 9000-level artificial intelligence, it's certainly enough to keep the bot out of trouble.

The end result is a semi-independent, mobile device that works via a computing "brain". It is a very simple robot, but that is a good thing because it offers potential and room for improvement.

How might you improve the robot's behaviour? Could you make it follow objects instead of just avoiding them? Can you think of any physical improvements to the robot that will make it useful?

Racer Bot

There are several professionally designed robot kits for the micro:bit, and we will use one of them to give an example of how MicroPython can be used to command and control quite a sophisticated robot.

The Bit:Bot by 4tronix (*http://4tronix.co.uk/bitbot*) is another trundlebot but tricked out with a large number of features: NeoPixels, a single pitch buzzer, line following sensors, light sensors, a proper edge connector for the micro:bit, and a battery holder with a power switch (see Figure 12-5). Assembly takes minutes.

Figure 12-5. The Bit:Bot by 4tronix

Control of the motors is more complicated than our homemade robot, since each motor has two connections: one for direction (forwards or backwards), the other for speed. The simplest way to make the motors move is to switch the pins to which they are connected to high:

```
pin8.write_digital(0)  # set direction to forwards
pin0.write_digital(1)  # set speed to full on
```

However, we probably want to change the speed of the motor, so instead we should replace the `write_digital` call on pin0 to its analog equivalent:

```
pin0.write_analog(511)
```

This is another appearance of PWM to drive something. Since the range for writing analog values is between 0 (off) to 1,023 (always on), then the preceding example sets the speed of the motor to half its potential full speed (since 511 is just under half of 1,023).

The 12 NeoPixels are connected to pin13 and use the standard `neopixel` module that comes with the micro:bit. The buzzer is monotonic (it only plays one note) and can be controlled with a digital signal via pin14. Reading digital values from pin 11 (left) and pin 5 (right) will indicate if a line is detected (allowing you to create a simple line following autopilot). The two light sensors are controlled by pin 16 (to select which one to use, 0 means left; 1 means right) and pin 2 (to read the analog value representing the amount of light detected).

It is possible to create a remote-controlled robot with two micro:bits connected using the `radio` module. This will require signalling between devices and a way to turn such interactions from the controlling micro:bit into outcomes on the remote controlled robot. In other words, there needs to be an intuitive user interface.

It should be possible to tilt the controlling micro:bit to indicate the direction of travel. For example, tilting the micro:bit forwards and left will cause the robot to move forwards and to the left. Furthermore, there should be some measurement of the degree of tilt so that the robot changes speed or angle of steering. One of the buttons should cause the buzzer to bleep like a sort of minimalist robot car horn, and the other could toggle the NeoPixels on and off. We'll set the line following and light sensors aside for the moment.

Assuming the control outlined, a protocol for sending actionable information is still needed. We need some way to send signals for the speed, steering, the buzzer, and NeoPixels; in other words, four pieces of information. The simplest and easiest-to-understand solution would be to send a string with the four values delineated by a token such as ":". If the place order of the values is pre-defined, the receiving robot can split the string at each instance of the ":" token and work with the resulting value. A message would look something like `speed:steer:buzzer:neopixel`, with each position being either an analog value (in the case of the speed and steer) or a digital value to indicate to sound the buzzer or toggle the NeoPixels. All that would remain is an agreement on the radio channel.

The following script for the controlling micro:bit implements all these features:

```python
from microbit import *
import radio


radio.config(channel=44)
radio.on()


# Defines the range of valid tilt from accelerometer readings.
max_tilt = 1000
min_tilt = 199


while True:
    # Grab the inputs.
    y = accelerometer.get_y()  # Forwards / backwards.
    x = accelerometer.get_x() # Left / right.
    a = button_a.was_pressed()  # Horn.
    b = button_b.was_pressed()  # Toggle lights.

    # Data from the controller to be sent to the vehicle.
    # [speed, steer, buzzer, neopixel]
    control_data = [0, 0, 0, 0]
    if x < -min_tilt and y < -min_tilt:
        # forwards left
        display.show(Image.ARROW_NW)
        control_data[0] = max(y, -max_tilt)
        control_data[1] = max(x, -max_tilt)
    elif x < -min_tilt and y > min_tilt:
        # backwards left
        display.show(Image.ARROW_SW)
        control_data[0] = min(y, max_tilt)
        control_data[1] = max(x, -max_tilt)
    elif x > min_tilt and y < -min_tilt:
        # forwards right
        display.show(Image.ARROW_NE)
        control_data[0] = max(y, -max_tilt)
        control_data[1] = min(x, max_tilt)
    elif x > min_tilt and y > min_tilt:
        # backwards right
        display.show(Image.ARROW_SE)
        control_data[0] = min(y, max_tilt)
        control_data[1] = min(x, max_tilt)
    elif y > min_tilt:
        # backwards
        display.show(Image.ARROW_S)
        control_data[0] = min(y, max_tilt)
```

```
    elif y < -min_tilt:
        # forwards
        display.show(Image.ARROW_N)
        control_data[0] = max(y, -max_tilt)
    if a:
        # Sound the buzzer
        control_data[2] = 1
    if b:
        # Toggle the NeoPixels
        control_data[3] = 1
    if any(control_data):
        msg = '{}:{}:{}:{}'.format(*control_data)
        radio.send(msg)
    else:
        display.clear()
    sleep(20)
```

The bulk of the script is in an event loop in which any inputs from the accelerometer
and buttons are read; and, depending on the readings from the accelerometer, the
speed and steer values are set. If a button is pressed for the buzzer or NeoPixels, this
is set as a flag. So the driver has feedback about what the micro:bit thinks it is doing
and the display is updated with arrows to indicate the direction. Finally, if there's any
control data to send, it is transmitted via the radio.

Things are simpler for the micro:bit controlling the robot: it needs to consume the
signal, decode it, turn the speed, and steer into signals to drive the motors and react
to buzzer or NeoPixel signals:

```
from microbit import *
import radio
import neopixel


display.show(Image.SKULL)  # Logo :-)

colour = (244, 0, 244)  # NeoPixel colour to use for lights.
np = neopixel.NeoPixel(pin13, 12)
lights = False

radio.config(channel=44)
radio.on()


def move(speed, steer):
    # Sensible defaults that mean "stop".
    forward = 0
    left = 0
    right = 0
    if speed > 0:
        # Moving forward.
        forward = 1
```

```
            left = 1000 - speed
            right = 1000 - speed
        elif speed < 0:
            # In reverse.
            left = 1000 + (-1000 - speed)
            right = 1000 + (-1000 - speed)
        if steer < 0:
            # To the right.
            right = min(1000, right + abs(steer))
            left = max(0, left - abs(steer))
        elif steer > 0:
            # To the left.
            left = min(1000, left + steer)
            right = max(0, right - steer)
        # Write to the motors.
        pin8.write_digital(forward)
        pin12.write_digital(forward)
        pin0.write_analog(left)
        pin1.write_analog(right)


while True:
    pin14.write_digital(0)  # Switch off the horn
    try:
        msg = radio.receive()
    except:
        msg = None  # Networks are not safe!
    if msg is not None:
        # Get data from the incoming message.
        speed, steer, horn, light = [int(val) for val in msg.split(':')]
        move(speed, steer)  # Move the robot.
        if horn:
            # Sound the horn
            pin14.write_digital(1)
        if light:
            # Toggle lights
            if lights:
                np.clear()
                lights = False
            else:
                lights = True
                for i in range(12):
                    np[i] = colour
                    np.show()
    else:
        # No message? Do nothing!
        move(0, 0)
    sleep(20)
```

A move utility function is defined to hide away the common implementation details of setting the pins to drive the motors. It takes the transmitted speed and steer values and ensures that the correct direction of rotation and level of speed is sent to the motors so that the robot moves forward and backwards or steers left or right.

The event loop switches off the horn (so it will only make a short bleep sound), listens out for incoming radio messages, and, if a message is received, reacts to the content therein.

If you have more than one robot and enough micro:bits, you could organise a race. However, to stop interference from the wrong controller, you should ensure that the radio channel is different between each pair of devices. Furthermore, the robot displays a logo, so it's probably a good idea to change that to a contestant number.

To end, I want to suggest a couple of enhancements to our robot to challenge your newfound robotics skills. If you press both buttons at the same time, it should be possible to toggle in and out of autopilot mode. The autopilot could work in two different ways: it could follow a line, or it could follow a light source. There are line- and light-detecting sensors on the chassis, so have a play and try to work out how to make it work.

## CHAPTER 13

Idiomatic MicroPython

The prior handful of chapters deal with many technical aspects of working with MicroPython: the modules you have available to you, how they interact with the hardware, and how various protocols can be used to make interesting and useful things happen. However, there has been no discussion of how to use such knowledge to create *good code*: code that is idiomatic MicroPython or, as some in the community like to say, Pythonic. To learn what this means, we need to take a step back and consider Python in terms of both language design and programming culture before looking into how best to write Pythonic MicroPython on highly constrained embedded devices.

Why is Python such a popular language? What motivates so many to contribute to the Python community? Why is Python widely used as a teaching language?

In the 1990s, Guido van Rossum, the creator of Python and the project's Benevolent Dictator for Life (BDFL), used Python as the basis for a project called "Computer Programming for Everybody: A Scouting Expedition for the Programmers of Tomorrow". The opening paragraphs of the project's proposal provide one clue to Python's popularity:

> In the seventies, Xerox PARC asked: "Can we have a computer on every desk?" We now know this is possible, but those computers haven't necessarily empowered their users. Today's computers are often inflexible: the average computer user can typically only change a limited set of options configurable via a "wizard" (a lofty word for a canned dialog), and is dependent on expert programmers for everything else.

> We ask a follow-up question: "What will happen if users can program their own computer?" We're looking forward to a future where every computer user will be able to "open the hood" of their computer and make improvements to the applications inside. We believe that this will eventually change the nature of software and software development tools fundamentally.

We compare mass ability to read and write software with mass literacy, and predict equally pervasive changes to society. Hardware is now sufficiently fast and cheap to make mass computer education possible: the next big change will happen when most computer users have the knowledge and power to create and modify software.

At the heart of Python's outlook is a spirit of empowerment. It's no wonder people like it. But Python also has a reputation for being a beautiful, expressive, and fun-to-use language. Why?

I believe the answer lies in a cultural artefact called the "Zen of Python". To read it, open a standard Python REPL and type `import this`:

```
>>> import this
The Zen of Python, by Tim Peters

Beautiful is better than ugly.
Explicit is better than implicit.
Simple is better than complex.
Complex is better than complicated.
Flat is better than nested.
Sparse is better than dense.
Readability counts.
Special cases aren't special enough to break the rules.
Although practicality beats purity.
Errors should never pass silently.
Unless explicitly silenced.
In the face of ambiguity, refuse the temptation to guess.
There should be one-- and preferably only one --obvious
way to do it.
Although that way may not be obvious at first unless
you're Dutch.
Now is better than never.
Although never is often better than *right* now.
If the implementation is hard to explain, it's a bad idea.
If the implementation is easy to explain, it may be a good idea.
Namespaces are one honking great idea--let's do more of those!
```

It defines what it is to be "Pythonic", that is, to write idiomatic Python code. Its author, Tim Peters, describes it as a document that "succinctly channels the BDFL's guiding principles for Python's design into 20 aphorisms, only 19 of which have been written down". Its humorously framed outlook displays a striving for simplicity, clarity, pragmatism and a sense of playful fun. This is quite different to the usual presentation of programming languages as complex, obscure, dense, and serious endeavours. Python favours a simple, elegant, and easy-to-comprehend solution to programming problems. As Alex Martelli puts it in his *Python Cookbook* (O'Reilly), "To describe something as clever is not considered a compliment in the Python culture."

Put simply, Python's focus on simplicity, clarity, pragmatism, and fun is appealing, and this attitude is brought to embedded development with MicroPython.

The Zen of MicroPython

There is a Zen of MicroPython, concocted as an Easter egg in the micro:bit port of MicroPython.[1] To read it, connect to the REPL and import this:

```
>>> import this
The Zen of MicroPython, by Nicholas H. Tollervey

Code,
Hack it,
Less is more,
Keep it simple,
Small is beautiful,

Be brave! Break things! Learn and have fun!
Express yourself with MicroPython.

Happy hacking! :-)
```

It is appropriately short and is a metaphor:[2] while there is humour, pearls of wisdom, and even a rhyme,[3] its most important attribute is brevity. If you look at the way it is written, there are also linguistic tricks that hint at the subtle and extraordinarily clever tricks MicroPython pulls to work in such constrained environments.[4]

The Zen of MicroPython is to express as much as possible with as little as possible, and this should be reflected in your code. Why? Because on some platforms, such as the micro:bit, you have but 16k RAM and a call stack of no greater depth than 5. Nevertheless, despite such extraordinarily constrained conditions, it is still possible to create a remarkable amount of functionality. In order to do so, you need to be aware of some of the tricks-of-the-trade when working with MicroPython. That's what the rest of this chapter will examine.

Memory

Memory is perhaps the biggest source of problems for new MicroPython programmers. There are two types of memory you should be aware of: *flash memory* (that is persistent and written when you "flash" MicroPython onto the device) and *RAM* (volatile memory). Flash memory is usually in the order of hundreds of kilobytes,[5]

---

1 There are a number of such Easter eggs in the micro:bit port of MicroPython. They're there to reward the curious, playful, and fearless explorers who may use the device. They're well hidden, but reading the output of the help() function will lead you down the rabbit hole, if you know what you're looking for.

2 I should know, I wrote it!

3 If you speak English with an Australian or British accent.

4 The first "verse" actually contains three verses if you read it in different directions.

5 Not to be confused with a filesystem on a USB flash drive.

whereas RAM can be as low as 8 kilobytes (although it is usually a lot more, depending on your device).

On some devices, such as the micro:bit, flash memory is limited, and there may be a maximum size for your script. For example, a script embedded directly into the MicroPython hex file may only be up to 2 kilobytes in size. You could store larger scripts on the (limited) filesystem (around 20 kilobytes). But it's important to understand that doing something as simple as removing comments, reducing indentation to only 2 characters (instead of the usual 4), or getting rid of blank lines may allow you to fit more code into the limited amount of flash memory.

In its standard configuration, MicroPython works in the same way as CPython. It parses your Python script, compiles it to bytecode, then runs it in a virtual machine (VM). These steps take up RAM in two important ways: when compiling Python to bytecode and when executing the bytecode with the MicroPython VM. There is a further issue called heap fragmentation that also has an adverse effect on memory usage.

When you import code, the MicroPython compiler uses memory to turn Python code into bytecode. Once compilation is finished, this memory becomes available to MicroPython. The resulting bytecode is stored in RAM. In some cases (such as when several modules have already been converted so their bytecode will take up RAM), the import statement will produce a memory exception since the compiler has run out of RAM.

Another cause for compilation memory errors is if a module creates global objects that require RAM at time of import (RAM that is unavailable for later compilation tasks). Avoid code that runs on import: it is better to have initialisation code that runs after all the required modules have been imported. This maximises the available RAM for the compilation step.

If RAM is still an issue during the compilation step, you have two further options, depending on the version of MicroPython you are using.

Frozen modules and frozen bytecode allow you to "bake in" code into the firmware image you flash onto the device. How this is achieved depends on the MicroPython port you're using, so it's best to consult the port-specific README for instructions. Some ports don't support this feature at all. At a high level, the steps are the same: put the target Python code into certain directories (depending on whether the code is to be a frozen module or frozen bytecode), build the firmware (although a specific command may be needed to freeze the code), and flash the resulting firmware onto the device. The end result is the code you've "frozen" into the firmware can be accessed with an import statement.

If re-compiling the MicroPython firmware is inappropriate, it is also possible to manually pre-compile bytecode on your PC and copy the resulting .mpy files over to the device's filesystem, thus mitigating the need for compilation. The MicroPython cross compiler can be found in the `mpy-cross` directory of the project's source code. The cross compiler runs on any Unix-like operating system and, in its simplest form, merely requires you to do this:

```
$ ./mpy-cross my_module.py
```

This results in a file called `my_module.mpy` that you should copy onto the device's filesystem. In your code, import it as you would any other Python module with `import my_module`.

Once the compilation step is completed, execution of the bytecode takes place and uses RAM. There are a number of ways in which you can reduce the footprint of RAM usage during the execution phase.

Some (but not all) ports of MicroPython have a `const` keyword that works in a similar way to `#define` in the C language. When your code is compiled to bytecode, the compiler will avoid using a lookup to the name of the constant by substituting its literal value. This saves bytecode and thus RAM. Usage is simple:

```
from micropython import const
ROWS = const(33)
_COLS = const(0x10)
a = ROWS
b = _COLS
```

The value passed into `const` must be anything that, at compile time, evaluates to an integer. In the example, the `ROWS` value will take up room in the globals dictionary because it will be available for other modules to import and use. However, because of the pre-pended underscore, the `_COLS` value is not available outside the current module and thus takes up no RAM.

Another useful technique is to save memory with constant data structures that never change during the execution of your code. If you are using frozen bytecode, you should consider data expressed as `bytes` objects. Since `bytes` objects are immutable, the compiler will ensure such objects remain in flash memory rather than getting copied over to RAM. The `ustruct` module built into MicroPython can be used to convert between `bytes` types and other Python types.

This trick works on other immutable values (such as strings, floats, integers, and complex numbers) and means they will be stored in flash memory for the same reason. When you assign such an immutable object, the value resides in flash, and only a reference to the location of the value takes up any RAM.

One might expect a tuple (an immutable data structure) of immutable values to benefit from the same trick. However, this is a future enhancement.

If you are used to writing Python in less constrained environments where it's fine to create lots of objects, you may need to rethink how you change your programming style to make the use of RAM more efficient.

For example, when concatenating strings, try to do it in such a way that it happens at compile time:

```
foobar1 = "foo" + "bar"  # bad
foobar2 = "foo" "bar"  # good
```

Both produce "foobar", but the first creates two string objects and allocates RAM for concatenation before producing a third object. The second just concatenates at compile time.

When strings need to be written to or read from a stream (such as a file), do it in a lazy manner. Rather than create a large string object that takes up RAM, work in a piecemeal fashion with smaller chunks. In a related manner, if you're reading data from, say, I²C, use a pre-allocated buffer rather than needlessly creating new objects as you loop over the data:

```
# Bad
while True:
    var = spi.read(100)
    # process data

# Good
buf = bytearray(100)
while True:
    spi.readinto(buf)
    # process data in buf
```

Rather than create a new buffer on each pass (as the bad snippet does), you should re-use a buffer, which is much faster and, as we shall see, helps to avoid memory fragmentation.

Another trick involves using the most memory-efficient representation of data. For example, an integer usually takes up 4 bytes, so if you can just use bytes for smaller numeric values, you'll save RAM. For example, (1, 2, 3) takes up more RAM than b'\1\2\3'. If this were in a frozen bytecode module, the bytes object would reside in the flash memory.

Memory fragmentation is an interesting problem you should be aware of. Imagine you create two objects called A and B. If A is reclaimed, but B remains at a higher memory address, then only objects no bigger than A will be able to reuse the space in the memory that is left over. If there's a lot of object creation and reclamation in your code, then there's a danger of such memory fragmentation occurring: despite there being substantial amounts of RAM available, there is not enough contiguous RAM to store some objects of a certain size. The result is a memory error.

This is why it's better to use a single object of fixed size, such as a reusable buffer. Furthermore, where such large, permanent buffers are needed, it's best to instantiate these early in the program before fragmentation can occur.

Under the hood, MicroPython's garbage collector is managing memory for you. There is a section of memory called the heap. When an object is created, it is stored on the heap. When it goes out of scope in your code (i.e., there are no more references to it to allow you to access it), then its chunk of the heap is reclaimed for other objects by the garbage collector (GC). This process happens automatically, so you mostly never have to worry about memory. However, you can access the garbage collector yourself via the gc module and force MicroPython to reclaim such redundant objects with the gc.collect function. If you need to call the garbage collector, you will get better performance if you do it a little and often.

The gc and micropython modules also contain functions that allow you to examine how memory is used for the purposes of debugging. For example, gc.mem_free and gc.mem_alloc tell you how much of the heap is free or allocated. The micropython.mem_info function will report a summary of memory usage if called with no arguments and will display a table of heap allocation if you pass in an arbitrary argument (such as 1):

```
>>> micropython.mem_info(1)
stack: 448 out of 15360
GC: total: 102080, used: 2832, free: 99248
 No. of 1-blocks: 45, 2-blocks: 14, max blk sz: 68, max free sz: 6194
GC memory layout; from 20003140:
00000: h=hhhBhhhhhBTBhhh==h=hh===hBhTh=hSBhTh=hh=hSBhTh=hh=hShhhh=Th==h
00400: =Bh=Bh=h=Shhh=======h==========================================
00800: =====================hB..h...h========h========h=....h=......
       (96 lines all free)
18c00: ..........................................
```

The table uses 10 symbols:

- . (free block)
- h (head block)
- = (tail block)
- m (marked head block)
- T (tuple)
- L (list)
- D (dict)
- F (float)
- B (byte code)
- M (module)

Each letter in the table is a single block of memory (16 bytes).[6]

The `micropython.alloc_emergency_exception_buf` is also an extraordinarily useful temporary aid when debugging your work. If you're connected to the REPL, in certain memory constrained conditions, if you encounter an exception you won't get a traceback containing details of the error. You'll just see the name of the exception. However, it's possible to tell MicroPython to pre-allocate some bytes of RAM as an emergency exception buffer, thus allowing you to see useful traceback information about these situations. This is very helpful when debugging; however, this feature shouldn't be considered appropriate under "normal" usage situations, since it takes up memory. A good way to use this feature is to put it at the start of your "main.py" script so that it'll be active for all subsequent code. A good size for the buffer is around 100 bytes:

```
import micropython

micropython.alloc_emergency_exception_buf(100)
```

One final debugging trick to share (originally invented by Carlos Pereira Atencio) is the pseudobreakpoint. In "regular" Python it's possible to attach a debugger to a Python process and tell it to stop at certain line numbers in the source code, thus allowing you to investigate the state of the program at that point in its execution. Unfortunately, there is no such debugger for MicroPython. However, by using an infinite loop immediately prior to the code of interest, MicroPython will patiently wait until the REPL is connected and you press CTRL-C. Simply use `while True:` `pass` in these situations. It's nowhere nearly as powerful as a real breakpoint, nor is it a universal replacement for a breakpoint, but it does give you something similar to a breakpoint's behaviour of pausing to allow you to interrogate the current context.

Performance

The microcontrollers upon which MicroPython runs are very slow compared to the CPUs of other sorts of devices. Sometimes this is a problem and you need to improve the performance of your code.

One obvious attribute of well-performing code is that the algorithm you're using is efficient. This is something *under your control* and not really in the scope of this book (i.e., it's up to you to be a competent programmer so that you research and implement an appropriately efficient algorithm for your task).

Another important aspect of your code is that it allocates and uses resources efficiently (as covered in the previous section on memory usage).

---

6 The minimum allocation unit for memory is 16 bytes.

A final, and too-obvious-to-mention-but-it's-always-missed-anyway aspect of your code is to check that it actually does what you expect it to.

Since attaching a debugger to a running MicroPython process on a board is out of the question, then connecting to the REPL, ensuring you make appropriate use of print, and watching the resulting messages scroll by is probably your best course of action to check the state of your running code. Remember, too, that if you interrupt the running program, all the objects will still be available for you to inspect interactively via the REPL.

Some boards that run MicroPython don't have floating-point hardware, so such operations are done in software (and are thus much slower). If this is the case, where possible, use integer operations, and only use floating-point arithmetic in parts of the code where performance isn't so important.

If after taking these steps you still encounter performance problems, you should profile your code. If your port of MicroPython has the time module, the following decorator function will measure the execution time of any method or function to which it is annotated:

```
def timed_function(f, *args, **kwargs):
    myname = str(f).split(' ')[1]
    def new_func(*args, **kwargs):
        t = time.ticks_us()
        result = f(*args, **kwargs)
        delta = time.ticks_diff(time.ticks_us(), t)
        print('Function {} Time = {:6.3f}ms'.format(myname, delta/1000))
        return result
    return new_func
```

Once you have identified problematic code, you have several options to improve performance (in addition to improving your Python code with the suggestions outlined). On some boards, you can instruct the compiler to emit ARM native opcodes rather than MicroPython bytecode. Most Python code can be used with this approach, although context managers and generators are not supported; and if raise is used, an argument must be supplied. To use this functionality, decorate the target function with micropython.native. Unfortunately, there is a trade-off: an increase in the size of compiled code (despite the end result running twice as fast as bytecode).

Depending on your board, in addition to bytecode or native opcodes, MicroPython can emit a third type of output: viper. Decorating with micropython.viper produces an optimised version of ARM native opcodes (such as improvements in integer arithmetic), although not all of Python's features are supported.

When available, these two more performant emitters are like a sort of manual JIT compiler (such as the one built into PyPy). Usually the JIT will analyse code and, when certain conditions are met, abandon Python bytecode and switch to native instructions, thus automatically improving performance. However, a JIT is complicated and therefore not desirable for underpowered microcontrollers. When available, such manual indications to produce more performant output is a "good enough" compromise.

Another option to improve performance is to use inline assembler code for microcontrollers with ARM's Thumb2 architecture. It is beyond the scope of this book to deal with the details of assembler language, but the following example demonstrates how to create a function with inline assembler and call it within your Python code:

```
@micropython.asm_thumb
def asm_add(r0, r1):
    add(r0, r0, r1)
```

The function must be decorated with `micropython.asm_thumb`. It's possible for such inline assembler functions to accept up to four arguments that must be named `r0`, `r1`, `r2`, and `r3`. The function defined in the previous example takes two arguments and adds them together in a way that is equivalent to `r0 = r0 + r1` in Python. By convention, because the result is put into register `r0`, that is what is returned.

A final option is to create your module in C, recompile MicroPython, and use it as a native module. MicroPython is written in ANSI C, modular, well tested, and has plenty of code examples for how to do this. A word of warning to those of you with C experience: the Zen of MicroPython applies here too. Coding in C for microcontrollers is quite different to C on regular devices. Conciseness, a regard for the limited resources (especially memory), and efficiency are paramount.

In general, it is worth re-emphasising that MicroPython is mostly just regular Python. On the whole, for short scripts, standard idiomatic Python (a la the Zen of Python) will work just fine. If you get memory errors or performance isn't what it should be, then follow the Zen of MicroPython and make use of the techniques and tricks outlined in this chapter. The worst thing that can possibly happen is you learn something new. Hopefully, you'll overcome your gremlin and end up fist-pumping the air while shouting "woohoo" when your special blinkenlight demo works as expected without causing a memory error or slowing down.

## CHAPTER 14

Next Steps

MicroPython is a relatively young project, yet it is gaining momentum all the time.

Ports targeting new hardware are in development. The implementation is improving. Thanks to the micro:bit, a huge number of educational resources already exist. Established players in the "maker" and electronics space are using and promoting it in their products (such as Adafruit). Conferences are creating interactive badges for attendees from microcontrollers that run MicroPython. Thanks to the European Space Agency (ESA), MicroPython may end up on payloads in space. Talks about MicroPython are appearing at PyCons all over the world. New libraries and code for interacting with all sorts of interesting peripherals are released every day.

This is an exciting time to get involved with the project and explore embedded programming with Python.

However, reading this book is but a first step. If you have a device, don't just let it sit in your desk drawer. Use it for a goofy weekend project, a work-related hack-day, or as the basis of a talk at your local user group. If you have kids, share the project with them and introduce them to programming (for example, create a motion detecting "parent trap" that sounds an alarm when you enter their bedroom). After just a couple of projects, you'll start to see the potential in making use of MicroPython and the devices upon which it runs in all sorts of interesting and previously unimagined projects.

The Community

Why is Python's community so important?

When you are part of a programming community, you become aware of the different sorts of skills your peers bring to bear. You notice where people share their projects, so you can learn from each other. You discover the common mistakes that people make. You may even collaborate with the people who build the library you're using in your project. Ultimately, you'll make friends and build a support network for those times when you need help. Crucially, it offers you an opportunity to help others, gain recognition, and contribute experience and resources back into the community. It's a virtuous circle.

The wider Python community has an excellent reputation for being a friendly group of people who value openness, actively engage in outreach (just look at all the educational projects in the Python community), and who organise some of the most interesting, diverse, and fun software conferences on the planet.

The Python community is well organised, having created the Python Software Foundation (PSF) as a rallying point for the community. It's a volunteer-led organization devoted to advancing open source technology related to the Python programming language. You can join and support the PSF in its mission or even take part as a volunteer. The PSF is also a grant-giving body which supports projects that promote Python. This is an important mechanism that facilitates community-led support and development. If you have an idea for a MicroPython workshop that needs funding, you should apply for a grant. The process is easy, and the grants working-group is responsive and helpful.

Going Deeper

If you are interested in contributing code to MicroPython, porting MicroPython to a new board, or creating libraries for MicroPython, your first stop needs to be the project's website (*http://micropython.org/*). You will find a flourishing message board, links to the source code, and details of current ports and ports in progress. There's also the MicroPython Lib (*https://github.com/micropython/micropython-lib*) repository of core libraries ported to MicroPython from "regular" Python.

If you want to chat with other MicroPython users, there's a #micropython IRC channel on Freenode and a microbit-community on Slack.

If you want to roll up your sleeves and get coding, you should read the MicroPython developer documentation (*https://micropython-dev-docs.readthedocs.io/en/latest/index.html*) that outlines the project's structure, coding conventions, and expectations when it comes to such things as testing.

Tutorials are springing up all over the place, as are projects that use MicroPython. Adafruit has a growing number of freely available MicroPython projects and tutorials (*https://learn.adafruit.com/category/micropython*). Hackaday (a popular hardware hacking, maker website) also has a growing number of community-sourced projects (*https://hackaday.io/search?term=MicroPython*) for you to learn from. In 2015, there was even a micro:bit world tour (*https://microworldtour.github.io/*) where members of the Python community took part in a sort of digital chain letter and cooked up all sorts of interesting and educational projects for MicroPython on the micro:bit.

So get stuck in and remember...

Code,
Hack it,
Less is more,
Keep it simple,
Small is beautiful,

Be brave! Break things! Learn and have fun!
Express yourself with MicroPython.

Happy hacking! :-)
    —The Zen of MicroPython

Index

About the Author

Nicholas H. Tollervey is a classically trained musician, philosophy graduate, teacher, writer, and software developer. He's just like this biography: concise, honest, and full of useful information.

He's @ntoll (*https://twitter.com/ntoll*) on Twitter and blogs at *http://ntoll.org/*.

Colophon

The animal on the cover of *Programming with MicroPython* is a luna moth caterpillar (*Actias luna*).

Luna moth caterpillars can be found throughout southern Canada, eastern United States, and northern Mexico. In cooler climates, it produces one generation a year and in warmer climates, it can produce up to three generations in a year. Females lay 400–600 eggs on the underside of leaves and they incubate for eight to thirteen days.

Newly hatched, the caterpillars feast on local sources such as hickory, butternut, and walnut trees. These trees produce a toxic defensive chemical called juglone to inhibit insects from dining on them. Lunas develop an enzyme that helps them tolerate juglone in their diets.

After about a month of filling up on these plants, the caterpillar builds a cocoon. The insect lives inside for about three weeks, then emerges as a moth. Its wing markings reveal a crescent in the eyespot resembling a crescent moon, hence its namesake. The adult doesn't have a mouth or a digestive system and survives on caterpillar fat stores for merely a week while it mates and lays eggs to begin the life cycle again.

Many of the animals on O'Reilly covers are endangered; all of them are important to the world. To learn more about how you can help, go to *animals.oreilly.com*.

The cover image is from *Animal Life In The Sea and On The Land*. The cover fonts are URW Typewriter and Guardian Sans. The text font is Adobe Minion Pro; the heading font is Adobe Myriad Condensed; and the code font is Dalton Maag's Ubuntu Mono.

Learn from experts.
Find the answers you need.

Sign up for a **10-day free trial** to get **unlimited access** to all of the content on Safari, including Learning Paths, interactive tutorials, and curated playlists that draw from thousands of ebooks and training videos on a wide range of topics, including data, design, DevOps, management, business—and much more.

Start your free trial at:

oreilly.com/safari

(No credit card required.)

Milton Keynes UK
Ingram Content Group UK Ltd.
UKHW012227020924
447784UK00007B/135